The Chemistry of
Polymers

The Chemistry of Polymers

Margaret Morris

Larsen & Keller
www.larsen-keller.com

The Chemistry of Polymers
Margaret Morris
ISBN: 978-1-64172-373-2 (Hardback)

⊟ Larsen & Keller

Published by Larsen and Keller Education,
5 Penn Plaza,
19th Floor,
New York, NY 10001, USA

Cataloging-in-Publication Data

The chemistry of polymers / Margaret Morris.
    p. cm.
Includes bibliographical references and index.
ISBN 978-1-64172-373-2
1. Polymers. 2. Polymerization. 3. Macromolecules.
4. Chemical reactions. I. Morris, Margaret.
QD381 .C44 2020
547.7--dc23

For more information regarding Larsen and Keller Education and its products, please visit the publisher's website www.larsen-keller.com

Table of Contents

Preface

This book is a culmination of my many years of practice in this field. I attribute the success of this book to my support group. I would like to thank my parents who have showered me with unconditional love and support and my peers and professors for their constant guidance.

The scientific field that is concerned with the chemical synthesis, structure, and the physical and chemical properties of polymers and macromolecules is known as polymer chemistry. Its principles and methods are also applicable in a variety of sub-disciplines of chemistry such as organic chemistry, physical chemistry and analytical chemistry. On the basis of their origin, polymers are subdivided into biopolymers and synthetic polymers. The functional and structural materials that make most of the organic matter in organisms are biopolymers. Synthetic polymers are the structural materials that are manifested in synthetic fibers, paints, building materials, furniture, plastics, mechanical parts and adhesives. This book is a compilation of chapters that discuss the most vital concepts in the field of polymer chemistry. Some of the diverse topics covered herein address the varied branches that fall under this category. Those in search of information to further their knowledge will be greatly assisted by this book.

The details of chapters are provided below for a progressive learning:

Chapter – What is Polymer Chemistry?

The macromolecule which is composed of many repeated subunits is known as a polymer. The sub-discipline of chemistry which deals with the structure, chemical synthesis, physical and chemical properties of polymers is referred to as polymer chemistry. This is an introductory chapter which will briefly introduce all the significant aspects of polymers and polymer chemistry.

Chapter – Organic Polymers

The polymers which contain carbon atoms are known as organic polymers. Some of the various types of organic polymers are polyvinyl ether, polypyrrole, poly(diiododiacetylene), polydioctylfluorene, polyorthoester, polyphenylene sulphide and polythiophene. This chapter has been carefully written to provide an easy understanding of these types of organic polymers.

Chapter – Inorganic Polymers

The polymers which do not contain carbon atoms are known as inorganic polymers. Some of the common examples of inorganic polymers are geopolymer, polystannane, polythiazyl and polysilazane. All these different types of inorganic polymers have been carefully analyzed in this chapter.

Chapter – Polymerization

The process of reaction between monomer molecules in a chemical reaction in order to form three-dimensional network or polymer chains is referred to as polymerization. The major types of polymerization are coordination polymerization, dispersion polymerization, step-growth polymerization, nitroxide-mediated radical polymerization, ring-opening polymerization, etc. The topics elaborated in this chapter will help in gaining a better perspective about these types of polymerization.

Chapter – Polymers Stereochemistry

The subdiscipline of chemistry which deals with the study of the relative spatial arrangement of atoms which form the structure of molecules and their manipulation is known as stereochemistry. This chapter closely examines the key concepts of stereochemistry of polymers to provide an extensive understanding of the subject.

Margaret Morris

1
What is Polymer Chemistry?

The macromolecule which is composed of many repeated subunits is known as a polymer. The sub-discipline of chemistry which deals with the structure, chemical synthesis, physical and chemical properties of polymers is referred to as polymer chemistry. This is an introductory chapter which will briefly introduce all the significant aspects of polymers and polymer chemistry.

Polymers

Polymer is any of a class of natural or synthetic substances composed of very large molecules, called macromolecules, that are multiples of simpler chemical units called monomers. Polymers make up many of the materials in living organisms, including, for example, proteins, cellulose, and nucleic acids. Moreover, they constitute the basis of such minerals as diamond, quartz, and feldspar and such man-made materials as concrete, glass, paper, plastics, and rubbers.

The word *polymer* designates an unspecified number of monomer units. When the number of monomers is very large, the compound is sometimes called a high polymer. Polymers are not restricted to monomers of the same chemical composition or molecular weight and structure. Some natural polymers are composed of one kind of monomer. Most natural and synthetic polymers, however, are made up of two or more different types of monomers; such polymers are known as copolymers.

Organic polymers play a crucial role in living things, providing basic structural materials and participating in vital life processes. For example, the solid parts of all plants are made up of polymers. These include cellulose, lignin, and various resins. Cellulose is a polysaccharide, a polymer that is composed of sugar molecules. Lignin consists of a complicated three-dimensional network of polymers. Wood resins are polymers of a simple hydrocarbon, isoprene. Another familiar isoprene polymer is rubber.

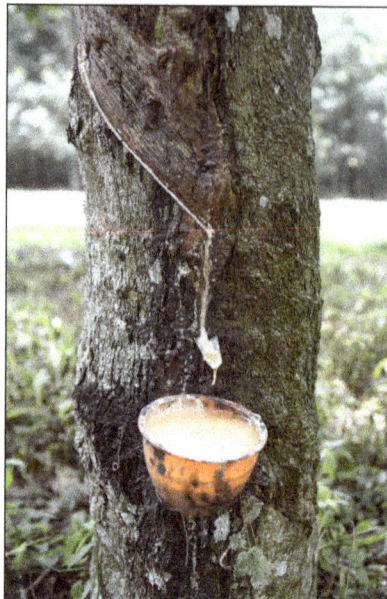

Natural Rubber: Latex tapped from a rubber tree (*Hevea brasiliensis*).

Other important natural polymers include the proteins, which are polymers of amino acids, and the nucleic acids, which are polymers of nucleotides—complex molecules composed of nitrogen-containing bases, sugars, and phosphoric acid. The nucleic acids carry genetic information in the cell. Starches, important sources of food energy derived from plants, are natural polymers composed of glucose.

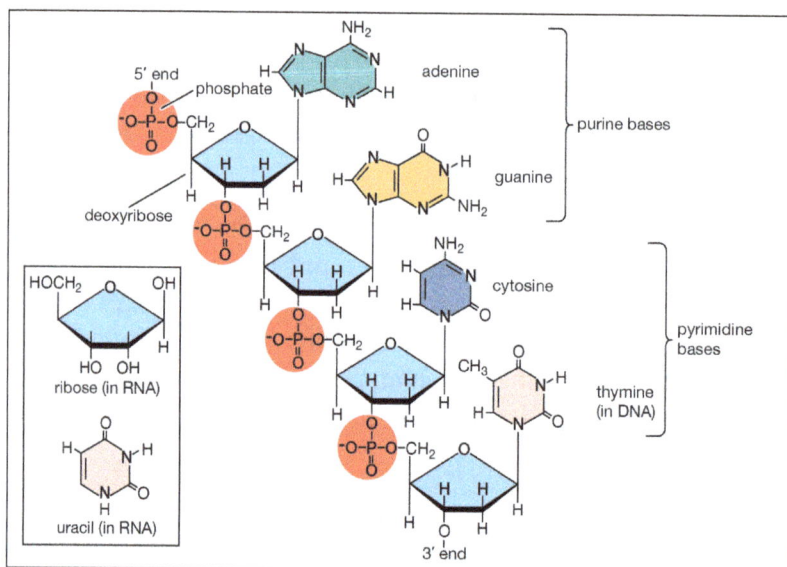

Portion of polynucleotide chain of deoxyribonucleic acid (DNA). The inset shows the corresponding pentose sugar and pyrimidine base in ribonucleic acid (RNA).

Many inorganic polymers also are found in nature, including diamond and graphite. Both are composed of carbon. In diamond, carbon atoms are linked in a three-dimensional network that gives the material its hardness. In graphite, used as a lubricant and in pencil "leads," the carbon atoms link in planes that can slide across one another.

Synthetic polymers are produced in different types of reactions. Many simple hydrocarbons, such as ethylene and propylene, can be transformed into polymers by adding one monomer after another to the growing chain. Polyethylene, composed of repeating ethylene monomers, is an addition polymer. It may have as many as 10,000 monomers joined in long coiled chains. Polyethylene is crystalline, translucent, and thermoplastic—i.e, it softens when heated. It is used for coatings, packaging, molded parts, and the manufacture of bottles and containers. Polypropylene is also crystalline and thermoplastic but is harder than polyethylene. Its molecules may consist of from 50,000 to 200,000 monomers. This compound is used in the textile industry and to make molded objects.

Other addition polymers include polybutadiene, polyisoprene, and polychloroprene, which are all important in the manufacture of synthetic rubbers. Some polymers, such as polystyrene, are glassy and transparent at room temperature, as well as being thermoplastic. Polystyrene can be coloured any shade and is used in the manufacture of toys and other plastic objects.

Polystyrene: Polystyrene packaging.

If one hydrogen atom in ethylene is replaced by a chlorine atom, vinyl chloride is produced. This polymerizes to polyvinyl chloride (PVC), a colourless, hard, tough, thermoplastic material that can be manufactured in a number of forms, including foams, films, and fibres. Vinyl acetate, produced by the reaction of ethylene and acetic acid, polymerizes to amorphous, soft resins used as coatings and adhesives. It copolymerizes with vinyl chloride to produce a large family of thermoplastic materials.

Many important polymers have oxygen or nitrogen atoms, along with those of carbon, in the backbone chain. Among such macromolecular materials with oxygen atoms are polyacetals. The simplest polyacetal is polyformaldehyde. It has a high melting point and is crystalline and resistant to abrasion and the action of solvents. Acetal resins are more like metal than are any other plastics and are used in the manufacture of machine parts such as gears and bearings.

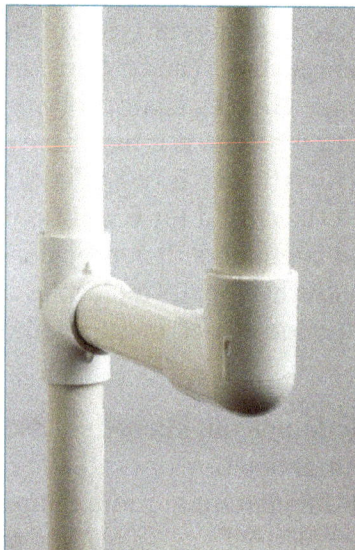

PVC piping: Polyvinyl chloride (PVC) pipes.

A linear polymer characterized by a repetition of ester groups along the backbone chain is called a polyester. Open-chain polyesters are colourless, crystalline, thermoplastic materials. Those with high molecular weights (10,000 to 15,000 molecules) are employed in the manufacture of films, molded objects, and fibres such as Dacron.

The polyamides include the naturally occurring proteins casein, found in milk, and zein, found in corn (maize), from which plastics, fibres, adhesives, and coatings are made. Among the synthetic polyamides are the urea-formaldehyde resins, which are thermosetting. They are used to produce molded objects and as adhesives and coatings for textiles and paper. Also important are the polyamide resins known as nylons. They are strong, resistant to heat and abrasion, noncombustible, and nontoxic, and they can be coloured. Their best-known use is as textile fibres, but they have many other applications.

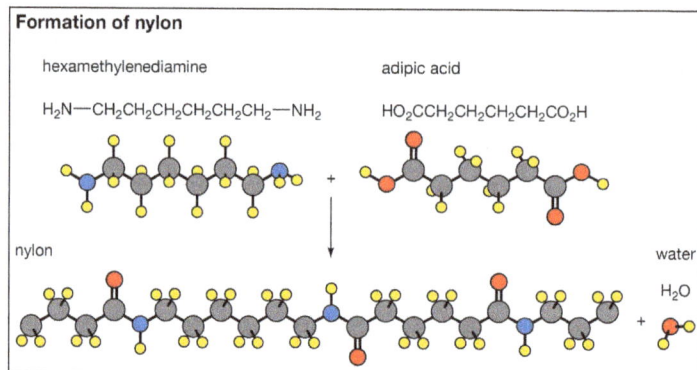

Nylon: The formation of nylon, a polymer.

Another important family of synthetic organic polymers is formed of linear repetitions of the urethane group. Polyurethanes are employed in making elastomeric fibres known as spandex and in the production of coating bases and soft and rigid foams.

A different class of polymers are the mixed organic-inorganic compounds. The most important representatives of this polymer family are the silicones. Their backbone consists of alternating silicon and oxygen atoms with organic groups attached to each of the silicon atoms. Silicones with low molecular weight are oils and greases. Higher-molecular-weight species are versatile elastic materials that remain soft and rubbery at very low temperatures. They are also relatively stable at high temperatures.

Caulk: Silicone caulk being dispensed from a caulking gun.

Polymer Chemistry

Polymer chemistry is the study of the synthesis, characterization and properties of polymer molecules or macromolecules, which are large molecules composed of repeating chemical subunits known as monomers.

Polymer chemistry is adept at producing a wide range of polymeric materials tailored to a variety of applications. Unfortunately, the human eye is unable to see atoms, so the beauty of the molecular architecture of these materials often goes unnoticed. Molecular modeling is able to solve this problem and enables the visualization of these complex structures along with an insight into their physical and mechanical properties. Polymer modeling is the counter part of the modeling used on a daily basis in the pharmaceutical industry. The vast sums of money and time that have to be invested to get a new drug to market have stimulated the field of pharmaceutical modeling, with protein homology modeling, protein-ligand docking, and computational design of combinatorial libraries commonplace these days. This is interesting as the average protein is a polymer of up to 20 monomers (amino acids), whereas the average synthetic polymer contains maybe one or two monomers. Hence, synthetic polymer modeling ought to be conceptually simpler than protein modeling but there has been less research into its applications. Polymer modeling is rapidly advancing to the stage where the modeling is quicker than the synthesis, and so computational screening of new potential polymers can be carried out prior to synthesis for desired properties. Therefore, we are embarking on the age of computer-aided polymer design.

There are a variety of ways to address the modeling of polymers, the most conceptually simple being to build a model of every atom in the polymer. This, however, is also the most computationally intensive method and requires a good knowledge of the system and software. A quicker method is to add parameters together to produce the desired property, from knowledge of the chemical groups in the polymer. This is termed QSPR (quantitative structure property relationships) and is the counter part of QSAR (quantitative structure activity relationships) in drug design. These QSPR methods are quick and easy to apply but lack the insight that the atomic detail can provide.

Classification of Polymers

Since Polymers are numerous in number with different behaviours and can be naturally found or synthetically created, they can be classified in various ways. The following below are some basic ways in which we classify polymers.

Classification Based on Source: The first classification of polymers is based on their source of origin, Let's take a look.

Natural Polymers

The easiest way to classify polymers is their source of origin. Natural polymers are polymers which occur in nature and are existing in natural sources like plants and animals. Some common examples are Proteins (which are found in humans and animals alike), Cellulose and Starch (which are found in plants) or Rubber (which we harvest from the latex of a tropical plant).

Synthetic Polymers

Synthetic polymers are polymers which humans can artificially create/synthesize in a lab. These are commercially produced by industries for human necessities. Some commonly produced polymers which we use day to day are Polyethylene (a mass-produced plastic which we use in packaging) or Nylon Fibers (commonly used in our clothes, fishing nets etc).

Semi-Synthetic Polymers

Semi-Synthetic polymers are polymers obtained by making modification in natural polymers artificially in a lab. These polymers formed by chemical reaction (in a controlled environment) and are of commercial importance. Example: Vulcanized Rubber (Sulphur is used in cross bonding the polymer chains found in natural rubber) Cellulose acetate (rayon) etc.

Classification Based on Structure of Polymers: Classification of polymers based on their structure can be of three types:

Linear Polymers

These polymers are similar in structure to a long straight chain which identical links connected to each other. The monomers in these are linked together to form a long chain. These polymers have high melting points and are of higher density. A common example of this is PVC (Poly-vinyl chloride). This polymer is largely used for making electric cables and pipes.

Branch Chain Polymers

As the title describes, the structure of these polymers is like branches originating at random points from a single linear chain. Monomers join together to form a long straight chain with some branched chains of different lengths. As a result of these branches, the polymers are not closely packed together. They are of low density having low melting points. Low-density polyethene (LDPE) used in plastic bags and general purpose containers is a common example

Crosslinked or Network Polymers

In this type of polymers, monomers are linked together to form a three-dimensional network. The monomers contain strong covalent bonds as they are composed of bi-functional and tri-functional in nature. These polymers are brittle and hard. Ex:- Bakelite (used in electrical insulators), Melamine etc.

Based on Mode of Polymerisation: Polymerization is the process by which monomer molecules are reacted together in a chemical reaction to form a polymer chain (or three-dimensional networks). Based on the type of polymerization, polymers can be classified as:

Addition Polymers

These type of polymers are formed by the repeated addition of monomer molecules. The polymer is formed by polymerization of monomers with double or triple bonds (unsaturated compounds). Note, in this process, there is no elimination of small molecules like water or alcohol etc (no by-product of the process). Addition polymers always have their empirical formulas same as their monomers. Example: ethene $n(CH_2=CH_2)$ to polyethene $-(CH_2-CH_2)_n-$.

Condensation Polymers

These polymers are formed by the combination of monomers, with the elimination of small molecules like water, alcohol etc. The monomers in these types of condensation reactions are bi-functional or tri-functional in nature. A common example is the polymerization of Hexamethylenediamine and adipic acid. to give Nylon – 66, where molecules of water are eliminated in the process.

Classification Based on Molecular Forces: *Intramolecular forces* are the *forces* that hold atoms together within a *molecule*. In Polymers, strong covalent bonds join atoms to each other in individual polymer molecules. *Intermolecular forces* (between the molecules) attract polymer molecules towards each other.

Note that the properties exhibited by solid materials like polymers depend largely on the strength of the forces between these molecules. Using this, Polymers can be classified into 4 types.

Elastomers

Elastomers are rubber-like solid polymers, that are elastic in nature. When we say elastic, we basically mean that the polymer can be easily stretched by applying a little force.

The most common example of this can be seen in rubber bands(or hair bands). Applying a little stress elongates the band. The polymer chains are held by the weakest intermolecular forces, hence allowing the polymer to be stretched. But as you notice removing that stress also results in the rubber band taking up its original form. This happens as we introduce crosslinks between the polymer chains which help it in retracting to its original position, and taking its original form. Our car tyres are made of Vulcanized rubber. This is when we introduce sulphur to cross bond the polymer chains.

Thermoplastics

Thermoplastic polymers are long-chain polymers in which inter-molecules forces (Van der Waal's forces) hold the polymer chains together. These polymers when heated are softened (thick fluid like) and hardened when they are allowed to cool down, forming a hard mass. They do not contain any cross bond and can easily be shaped by heating and using moulds. A common example is Polystyrene or PVC (which is used in making pipes).

Thermosetting

Thermosetting plastics are polymers which are semi-fluid in nature with low molecular masses. When heated, they start cross-linking between polymer chains, hence becoming hard and infusible. They form a three-dimensional structure on the application of heat. This reaction is irreversible in nature. The most common example of a thermosetting polymer is that of Bakelite, which is used in making electrical insulation.

Fibres

In the classification of polymers, these are a class of polymers which are a thread like in nature, and can easily be woven. They have strong inter-molecules forces between the chains giving them less elasticity and high tensile strength. The intermolecular forces may be hydrogen bonds or dipole-dipole interaction. Fibres have sharp and high melting points. A common example is that of Nylon-66, which is used in carpets and apparels.

The above was the general ways to classify polymers. Another category of polymers is that of Biopolymers. Biopolymers are polymers which are obtained from living organisms. They are biodegradable and have a very well defined structure. Various biomolecules like carbohydrates and proteins are a part of the category.

Structure of Polymers

Polymers are composed of non-metallic elements, found at the upper right corner of the periodic table. Carbon is the most common element in polymers. The chemical bonds in polymers are also different than those found in metals and ceramics.

Covalent Bonds

Non-metallic elements have a high number of valence electrons (four or more) and prefer to gain electrons, not lose them, in chemical reactions. They often form anions. In a compound of only nonmetals there are no elements willing to become cations, so ionic bonds are not possible. Instead, two nonmetallic atoms can share valence electrons with each other. This type of electron sharing, called covalent bonding, keeps the shared electrons close to both atomic nuclei. One pair of shared electrons makes one covalent bond. A molecule is a group of atoms held together by covalent bonds.

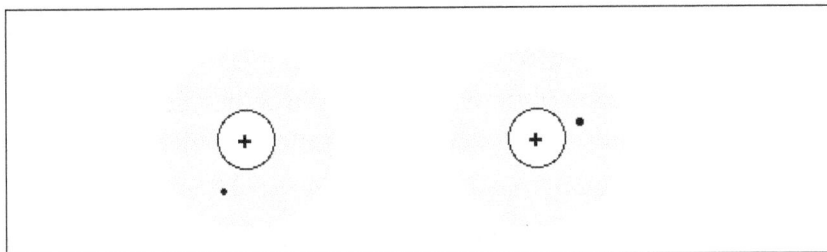

Each hydrogen atom has one proton in its nucleus and one electron.
The electron moves randomly in a spherical space around the nucles.

This type of bonding contrasts with metallic bonding, in which valence electrons are not associated with a particular nucleus, and move easily throughout a sample.

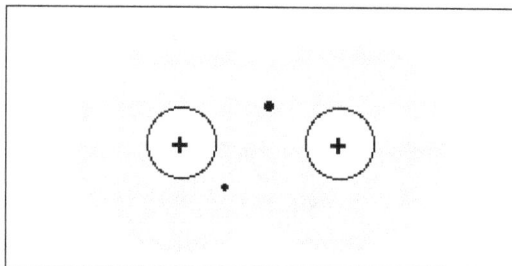

When a covalent bond forms, two hydrogen nuclei move close together.
Although the electrons still move around the molecule, ususally they are near both nuclei.

To determine how many covalent bonds can be formed between atoms, first the number of valence electrons must be counted. This can be determined by using a periodic table. The group number matches the number of valence electrons.

Example: Carbon is element 6. It is found in Group IV so has four valence electrons. Oxygen is element 8. It is found in Group VI so has six valence electrons. The molecule carbon dioxide has the chemical formula CO_2.

4 electrons from C + 2(6) electrons from O = 16 valence electrons

Lewis Structures

Rather than writing a sentence for the number of valence electrons on an atom, it can be more useful to draw a picture containing this information. A Lewis structure for an atom starts with a chemical symbol, with a dot added for each valence electron.

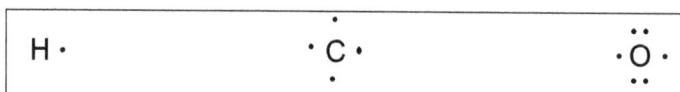

Since the highest possible number of valence electrons is eight (for noble gases) the dots representing valence electrons are traditionally arranged on four sides of the symbol, with at most two electrons on each side.

Experiments have shown that most nonmetallic nuclei are satisfied when they are near eight valence electrons. That is known as the octet rule. Carbon has four valence electrons, and needs to find four more to share. Oxygen has six valence electrons, so it only needs two more. Hydrogen is an exception to the octet rule; its nearest noble gas, helium, has only two electrons. Hydrogen nuclei form molecules with two nearby electrons, a duet rule.

To show a covalent bond, two chemical symbols are put near each other with two dots, representing a pair of electrons, between them. For example, a water molecule has one oxygen atom covalently bound to two hydrogen atoms.

water $:\overset{..}{\underset{..}{O}}:H$
                      H

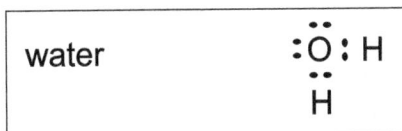

Nuclei do not have to share all of their valence electrons. Note that two pairs of the oxygen's valence electrons are not shared with any other atoms. Those electrons are called lone pairs or nonbonded pairs and can influence the chemical properties of a molecule.

Because dots can be difficult to see, it is common to draw a line segment for each bond (two electrons). The Lewis structure of a water molecule then looks like

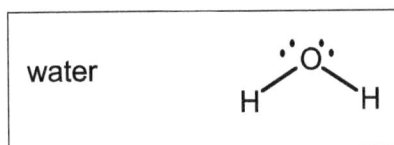

water $\overset{..}{\underset{}{O}}$
              H H

Hydrogen peroxide is a different compond of hydrogen and oxygen, with chemical formula H_2O_2. Since hydrogen atoms only need a duet of electrons they are found at the outside of the molecule; the two oxygen atoms need to be in the center where they can form more bonds. In this textbook that structural information will be given by underlining the central atom(s) in the chemical formula: $H_2\underline{O}_2$.

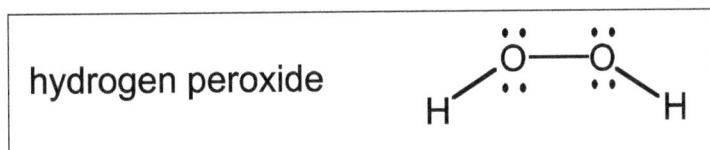

hydrogen peroxide $\overset{..}{\underset{..}{O}}—\overset{..}{\underset{..}{O}}$
                          H H

Sometimes nuclei will need to share more than one pair of electrons to achieve an octet with the available valence electrons. One shared pair of electrons is a single bond. Two shared pairs (four electrons) makes a double bond, and three shared pairs (six electrons) makes a triple bond. The more electrons that are shared, the stronger the bond will be. The neighboring elements carbon, nitrogen and oxygen commonly use double and triple bonds. For example, both nitrogen and oxygen are usually found as diatomic gases, N_2 and O_2. The nitrogen molecule has $2 \times 5 = 10$ valence electrons, and oxygen molecule has $2 \times 6 = 12$. Nitrogen needs a triple bond to achieve octets for each atom, but a double bond is sufficient for the oxygen molecule.

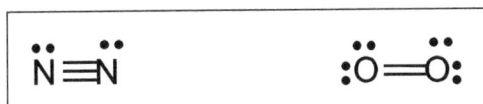

$\overset{..}{N}\equiv\overset{..}{N}$ $:\overset{..}{O}=\overset{..}{O}:$

The type of covalent bond affects the shape of a molecule. The nuclei move closer together if they share more electrons. This means that a triple bond is shorter than a double bond, which is shorter than a single bond. The bond angles are also very specific in a covalently bond molecule. The shared electrons want to be near the two positively charged nuclei, but try to stay away from negatively charged lone pairs.

This structure is quite different than that created by metallic bonds, which do not have a particular orientation. Covalent molecules can flex a bit under stress but prefer to "bounce back" to their original positions.

Although a Lewis structure is a good way to show covalent bonds between atoms, it is not as effective at showing a molecule's three dimensional shape. Chemists use model kits and chemical graphics programs to visualize the positions of atoms in molecules.

Functional Groups in Polymers

Carbon is the most important element in polymers. Because it starts with only four valence electrons, and wants to share four more, carbon forms a wide variety of covalent bonds. Most importantly, carbon forms strong bonds with itself. Long, strong chains or nets made of thousands of carbon atoms form the backbone of a polymer.

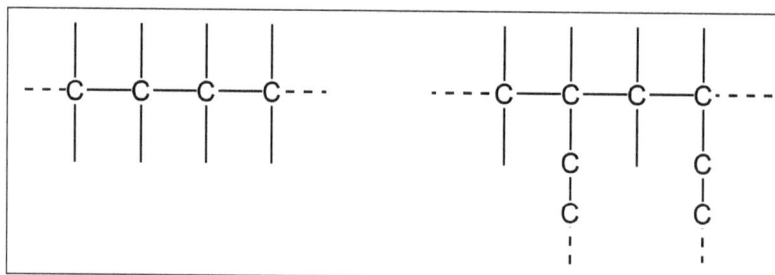

Carbon Backbones

Polyethylene is the simplest polymer. In addition to the carbon backbone, only hydrogen atoms are used to achieve four covalent bonds per carbon atom.

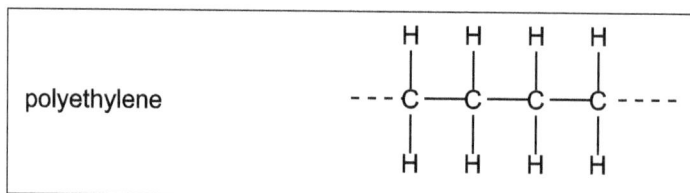

Although silicon is in the same group as carbon, it does not form strong bonds with itself. Silicones, long chains of alternating silicon and oxygen atoms, can be synthesized.

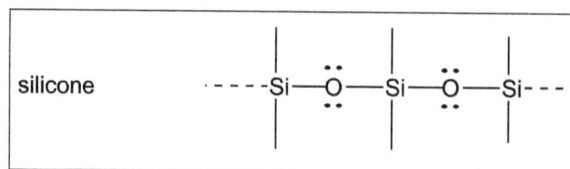

Many different nonmetal atoms could be covalently attached to a polymer backbone.

Groups of atoms that contribute something besides C-C and C-H bonds are called functional groups. They affect the chemical and physical properties of a polymer.

Examples of Functional Groups

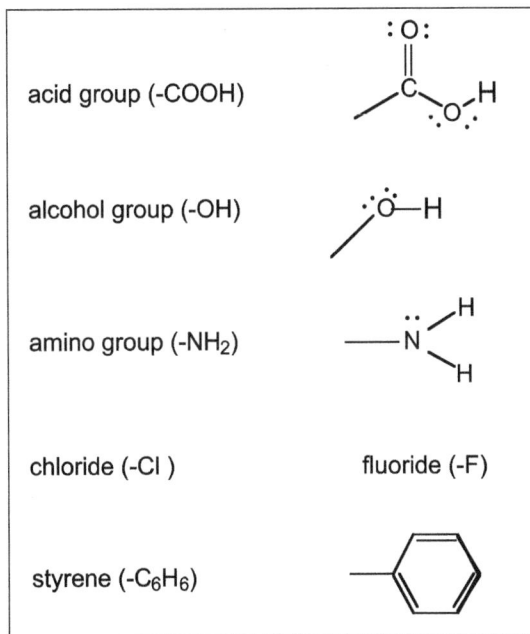

acid group (-COOH)	
alcohol group (-OH)	
amino group (-NH$_2$)	
chloride (-Cl)	fluoride (-F)
styrene (-C$_6$H$_6$)	

The primary experimental method used to identify functional groups in polymers is Infrared Spectroscopy (IR).

Skeleton Structures

Simplified or "skeleton" structures can be used to emphasize the functional groups. Carbon-carbon bonds of the framework are represented by line segments. Each vertex is the location of a carbon atom. Most hydrogen atoms and all lone pairs are omitted. This type of diagram deemphasizes the hydrocarbon skeleton; since it is so strongly bonded as to be unreactive, it does not affect the chemical properties of the polymer.

Polyethylene is the simplest polymer. Since it has no functional groups, the skeleton structure of a polyethylene fragment looks like it does not have any atoms! (Remember that a real polyethylene molecule is more often 100 or 1000 atoms long.)

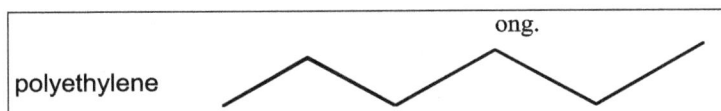

	ong.
polyethylene	

It is possible to figure out the missing information. There should be a carbon atom at the end of each line segment; six are needed, connected by five single bonds.

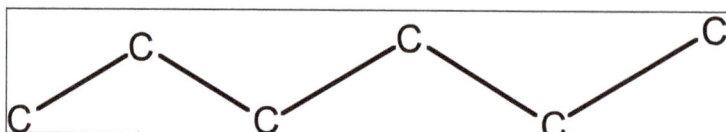

Since each carbon atom must have four bonds in a molecule, there must be missing bonds to hydrogen atoms. For the carbon atoms on the ends of the molecule, adding three C-H bonds to each will achieve octets. Two C-H bonds should be added to each of the inner carbons.

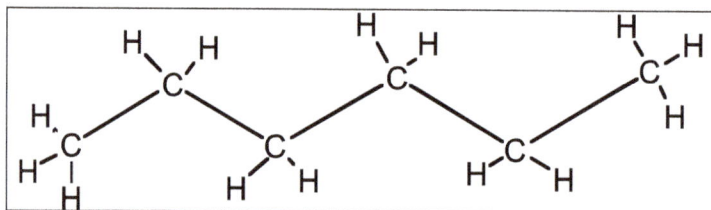

Complete Lewis structure for polyethylene fragment.

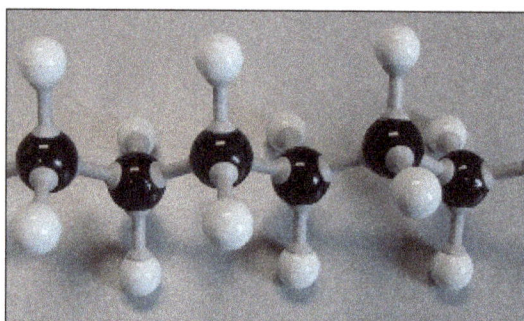

Three dimensional model of polyethylene.

When functional groups are added to a simplified backbone it is easy to notice the change in structure. Polyfluoroethylene, often sold as Teflon, is similar in structure to polyethylene except that all the hydrogen is replaced with fluorine. It is a very slippery polymer.

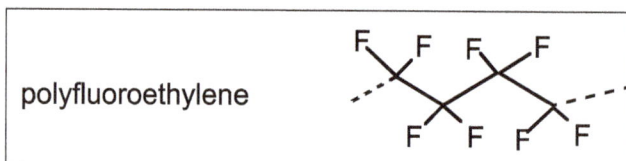

Amino acids, the monomers that build proteins, contain amino groups and acid groups, separated by one carbon. In the model nitrogen atoms are blue and oxygen atoms are red. The carbon between the amino group and the acid group always has one hydrogen on it (pointing up in the model) but the fourth group is variable. The symbol R is used when the exact identity of the group is not important. More than twenty different groups, as simple as a single hydrogen atom, are found on amino acids in nature.

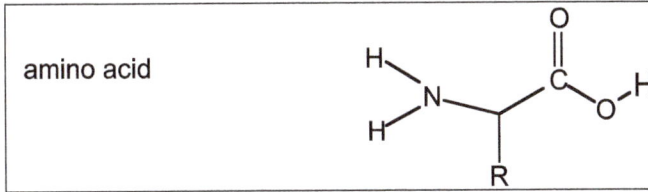

The photo shows a model of alanine, which has a -CH$_3$ group.

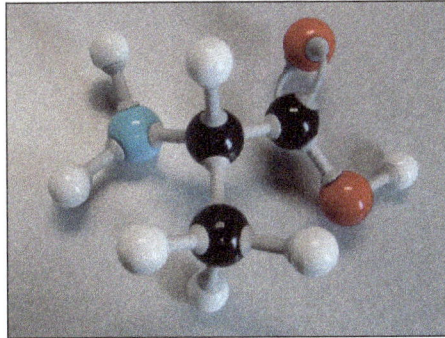

three dimensional model of amino acid alanine.

Polysaccharides are sugar polymers. Cellulose (found in wood, cabbages, cotton, and linen) is composed of long chains of sugar rings. They are covered with alcohol groups.

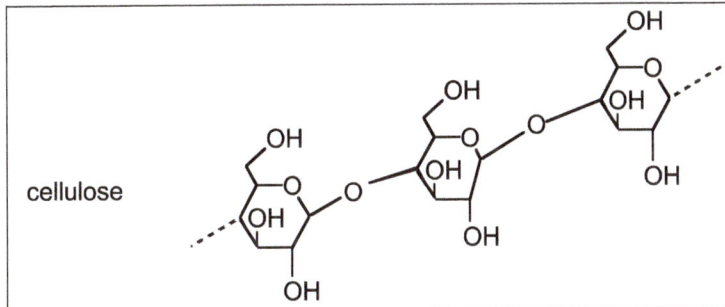

Starch contains the same functional groups, but the sugar rings are connected at different angles. The structural change makes it possible for the human digestive system to digest the polymer into sugar. Starch polymers are often branched.

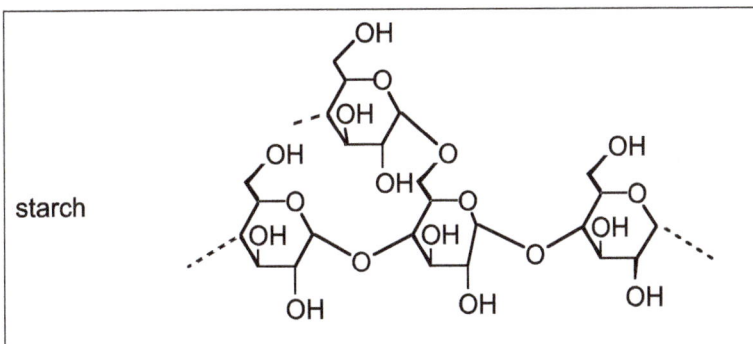

Intermolecular Forces

A molecule is a group of atoms connected by covalent bonds. Chemical reactions are required to form or break covalent bonds. Weaker attractions often form between molecules, encouraging them to stick together in groups. The weaker attractions are called secondary bonds or intermolecular forces. These can be overcome by adding heat or dissolving in a liquid. The functional groups on a polymer determine the type(s) and strength of its secondary bonds.

Polar Interactions

The valence electrons moving around a molecule may not be symmetrically distributed. The nonmetallic elements closest to the right top corner of the periodic table - nitrogen, oxygen, fluorine and chlorine - tend to shift shared electrons away from carbon and hydrogen. When there is a functional group with one of those elements, it has a slight negative charge and the rest of the molecule (carbon and hydrogen) is slightly positive. The molecule is polarized (or polar, for short). Its positive sections are attracted to negative sections of neighboring polymers.

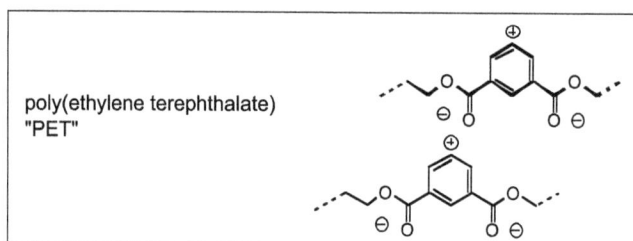

poly(ethylene terephthalate)
"PET"

Poly(ethylene terephthalate) or PET, a polymer used to make bottles for carbonated beverages, has oxygen-containing functional groups that make it polar. Protein and cellulose chains are also polar.

Polyfluoroethylene is nonpolar (not polar) because it is completely covered with fluorine atoms; there is no exposed positive section to interact with a neighboring molecule's negative section.

Positive and negative charges can be localized on a covalent molecule since they have no path for conduction of electrons. The carbon atoms in the backbone always follow the octet rule with four covalent bonds, so can't pass extra electrons along the chain. If polymer fibers are rubbed together they can build up a static electricity charge.

Hydrogen Bonds

Molecules with either -N-H or -O-H groups will form strong secondary bonds. This phenomenon is responsible for the relatively high boiling point of water, and for the fact that its solid form (ice) is less dense than its liquid form. Polymers with hydrogen-bonding groups will soak up water.

Fabric softeners are added to laundry to change the properties of cotton and linen fabrics. The fabric softener molecules have one end that binds to OH groups on cellulose. The other end of the fabric softener is a long, nonpolar chain. This exposed end feels smooth and slippery. Softened fabrics are less able to build up a static charge. However softened fabrics will not absorb as much water. This is an issue for the performance of cotton towels.

Nonpolar Interactions

As valence electrons move around the nuclei in a nonpolar polymer, like polyethylene or polyfluoroethylene, they can become temporarily imbalanced. For a brief moment of time one part of a molecule would be negative, another part positive; it is temporarily polar. These occasional imbalances are enough to allow nonpolar molecules to attract each other, but the interaction is much weaker than that observed for polar or hydrogen bonding polymers.

Properties and uses of Polymers

A polymer is a large molecule or a macromolecule which essentially is a combination of many subunits.. From the strand of our DNA which is a naturally occurring biopolymer to polypropylene which is used throughout the world as plastic.

Polymers may be naturally found in plants and animals (natural polymers) or may be man-made (synthetic polymers). Different polymers have a number of unique physical and chemical properties due to which they find usage in everyday life.

Polymers are all created by the process of polymerization wherein their constituent elements called monomers, are reacted together to form polymer chains i.e 3-dimensional networks forming the polymer bonds.

The type of polymerization mechanism used depends on the type of functional groups attached to the reactants. In biological contexts, almost all macromolecules are either completely polymeric or are made up of large polymeric chains.

Classification of Polymers

Polymers cannot be classified under one category because of their complex structures, different behaviours, and vast applications. We can, therefore, classify polymers based on the following considerations.

Classification of Polymers based on the Source of Availability

There are three types of classification under this category, namely, Natural, Synthetic, and Semi-synthetic Polymers.

Natural Polymers: They occur naturally and are found in plants and animals. For example proteins, starch, cellulose, and rubber. To add up, we also have biodegradable polymers which are called biopolymers.

Semi-synthetic Polymers: They are derived from naturally occurring polymers and undergo further chemical modification. For example, cellulose nitrate, cellulose acetate.

Synthetic Polymers: These are man-made polymers. Plastic is the most common and widely used synthetic polymer. It is used in industries and various dairy products. For example, nylon-6, 6, polyether's etc.

Classification of Polymers based on the Structure of the Monomer Chain

This category has the following classifications:

Linear Polymers: The structure of polymers containing long and straight chains fall into this category. PVC, i.e. poly-vinyl chloride is largely used for making pipes and electric cables is an example of a linear polymer.

Branched-chain Polymers: When linear chains of a polymer form branches, then, such polymers are categorized as branched chain polymers. For example, Low-density polythene.

Cross-linked Polymers: They are composed of bifunctional and trifunctional monomers. They have a stronger covalent bond in comparison to other linear polymers. Bakelite and melamine are examples in this category.

Other Ways to Classify Polymers

Classification Based on Polymerization

- Addition Polymerization: Example, poly ethane, Teflon, Polyvinyl chloride (PVC).

- Condensation Polymerization: Example, Nylon -6, 6, perylene, polyesters.

Classification Based on Monomers

- Homomer: In this type, a single type of monomer unit is present. For example, Polyethene.

- Heteropolymer or co-polymer: It consists of different type of monomer units. For example, nylon -6, 6.

Classification Based on Molecular Forces

- Elastomers: These are rubber-like solids weak interaction forces are present. For example, Rubber.

- Fibres: Strong, tough, high tensile strength and strong forces of interaction are present. For example, nylon -6, 6.

- Thermoplastics: These have intermediate forces of attraction. For example, polyvinyl chloride.

- Thermosetting polymers: These polymers greatly improve the material's mechanical properties. It provides enhanced chemical and heat resistance. For example, phenolics, epoxies, and silicones.

Structure of Polymers

Most of the polymers around us are made up of a hydrocarbon backbone. A Hydrocarbon backbone being a long chain of linked carbon and hydrogen atoms, possible due to the tetravalent nature of carbon.

A few examples of a hydrocarbon backbone polymer are polypropylene, polybutylene, polystyrene. Also, there are polymers which instead of carbon have other elements in its backbone. For example, Nylon, which contains nitrogen atoms in the repeated unit backbone.

Types of Polymers

On the basis of the type of the backbone chain, polymers can be divided into:

- Organic Polymers: Carbon backbone.

- Inorganic Polymers: Backbone constituted by elements other than carbon.

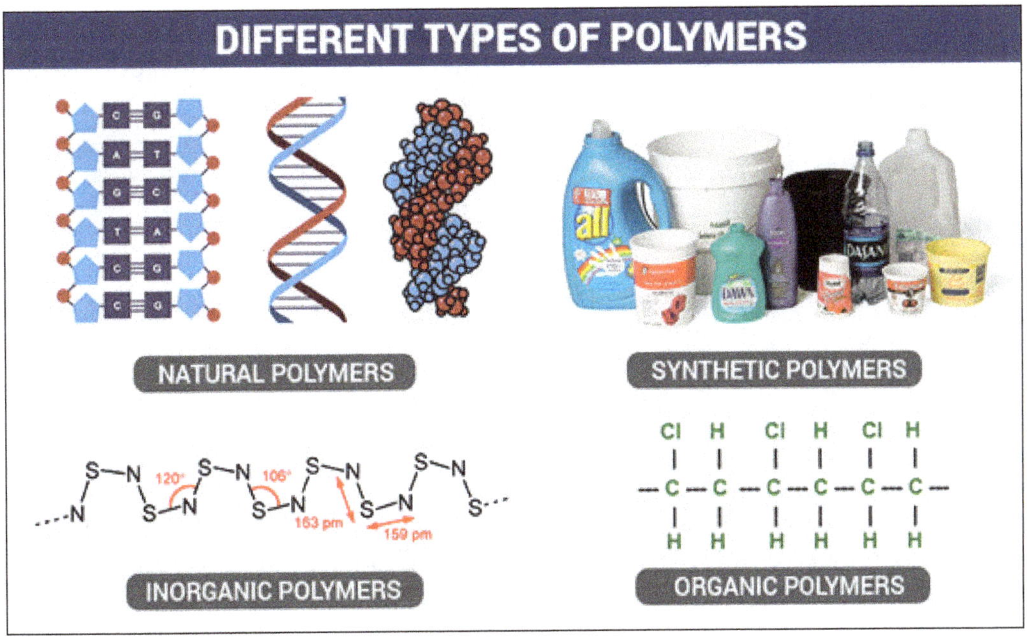

DIFFERENT TYPES OF POLYMERS

NATURAL POLYMERS

SYNTHETIC POLYMERS

INORGANIC POLYMERS

ORGANIC POLYMERS

On the basis of their synthesis:

- Natural Polymers

- Synthetic Polymers

Biodegradable Polymers

The polymers which are degraded and decayed by microorganisms like bacteria are known as biodegradable polymers. These types of polymers are used in surgical bandages, capsule coatings and in surgery. For example, Poly hydroxybutyrate co vel [PHBV].

High-temperature Polymers

These polymers are stable at high temperatures. Due to their high molecular weight, these are not destroyed even at very high temperatures. They are extensively used in the healthcare industries, for making sterilization equipment and in the manufacturing of heat and shock-resistant objects.

Few of the Important Polymers are:

Polypropylene: It is a type of polymer that softens beyond a specific temperature allowing it to be moulded and on cooling it solidifies. Due to its ability to be easily moulded into various shapes, it has a lot of applications.

A few of which are in stationary equipment's, automotive components, reusable containers speakers and much more. Due to its relatively low energy surface, the polymer is fused with the welding process and not using glue.

Polyethene: It is the most common type of plastic found around us. Mostly used in packaging from plastic bags to plastic bottles. There are different types of polyethene but their common formula being $(C_2H_4)_n$.

Properties of Polymers

Physical Properties:

- As chain length and cross-linking increases the tensile strength of the polymer increases.

- Polymers do not melt, they change state from crystalline to semi-crystalline.

Chemical Properties:

- Compared to conventional molecules with different side molecules, the polymer is enabled with hydrogen bonding and ionic bonding resulting in better cross-linking strength.

- Dipole-dipole bonding side chains enable the polymer for high flexibility.

- Polymers with Van der Waals forces linking chains are known to be weak, but give the polymer a low melting point.

Optical Properties:

- Due to their ability to change their refractive index with temperature as in the case of PMMA and HEMA: MMA, they are used in lasers for applications in spectroscopy and analytical applications.

Some Polymers and their Monomers:

- Polypropene, also known as polypropylene, is made up of monomer propene.

- Polystyrene is an aromatic polymer, naturally transparent, made up of monomer styrene.

- Polyvinyl chloride (PVC) is a plastic polymer made of monomer vinyl chloride.

- The urea-formaldehyde resin is a non-transparent plastic obtained by heating formaldehyde and urea.

- Glyptal is made up of monomers ethylene glycol and phthalic acid.

- Bakelite or polyoxybenzylmethylenglycolanhydride is a plastic which is made up of monomers phenol and aldehyde.

Types of Polymerization Reactions

Addition Polymerization: This is also called as chain growth polymerization. In this, small monomer units joined to form a giant polymer. In each step length of chain increases. For example, Polymerization of ethane in the presence of Peroxides.

Condensation Polymerization: In this type small molecules like H_2O, CO, NH_3 are eliminated during polymerization (step growth polymerization). Generally, organic compounds containing bifunctional groups such as idols, -dials, diamines, dicarboxylic acids undergo this type of polymerization reaction. For example, Preparation of nylon -6, 6.

1. What is Copolymerization?

In this process, two different monomers joined to form a polymer. Synthetic rubbers are prepared by this polymerization. For example, BUNA – S.

2. How to Calculate Molecular Mass of Polymers?

There are two types of average molecular masses of Polymers:

- Number Average Molecular Masses.

- Weight Average Molecular Mass.

Number Average Molecular Masses: If N_1, N_2, N_3.... are the number of macromolecular with molecular masses. M_1, M_2, M_3..... Respectively then the number average molecular masses of the polymer is given by:

$$\bar{M}_n = \frac{N_1M_1 + N_2M_2 + N_3M_3 + ...\Sigma N_iM_i}{N_1 + N_2 + N_3 + ...\Sigma N_i}$$

The number average molecular mass $\bar{M}_n$ is determined by colligative properties such as Osmotic Pressure.

Weight Average Molecular Mass: If m_1, m_2, m_3.... Are the masses of a macromolecule with molecular masses M_1, M_2, M_3... respectively, Then weight average molecular mass of the polymer is given by:

$$\bar{M}_\omega = \frac{m_1M_1 + m_2M_2 + m_3M_3 + ...}{m_1 + m_2 + m_3}$$

$$\Rightarrow = \frac{\Sigma miMi}{\Sigma mi}$$

$$\Rightarrow \bar{M}_\omega = \frac{\Sigma NiMi \times Mi}{\Sigma NiMi}$$

$$\Rightarrow \bar{M}_\omega = \frac{\Sigma NiMi^2}{\Sigma NiMi}$$

Polydispersive index: It is the ratio of weight average molecular mass and number average molecular mass of Polymers.

$$PDI = \frac{\bar{M}\omega}{\bar{M}n} \text{ For natural polymers PDI} = 1.$$

Uses of Polymers:

- Polypropene finds usage in a broad range of industries such as textiles, packaging, stationery, plastics, aircraft, construction, rope, toys, etc.

- Polystyrene is one of the most common plastic, actively used in the packaging industry. Bottles, toys, containers, trays, disposable glasses and plates, tv cabinets and lids are some of the daily-used products made up of polystyrene. It is also used as an insulator.

- The most important use of polyvinyl chloride is the manufacture of sewage pipes. It is also used as an insulator in the electric cables.

- Polyvinyl chloride is used in clothing and furniture and has recently become popular for the construction of doors and windows as well. It is also used in vinyl flooring.

- Urea-formaldehyde resins are used for making adhesives, moulds, laminated sheets, unbreakable containers, etc.

- Glyptal is used for making paints, coatings, and lacquers.

- Bakelite is used for making electrical switches, kitchen products, toys, jewellery, firearms, insulators, computer discs, etc.

Molecular Weight, Size and Branching of Polymers

Polymers are extensively used throughout industry, whether as naturally occurring materials like starch and cellulose, or synthetic commodities like nylon, polystyrene and polyethylene. Polymer functionality is defined by its molecular weight (MW), molecular size and structure, MW distribution and the degree of cross linking or chain branching. Hence, the polymer manufacturer needs to implement efficient and effective methods for determining these properties.

The word "polymer" has been derived from the Greek words 'poly' and 'meros', implying many parts, and refers to a characterizing feature of polymeric materials – their chain like structure. This structure is formed by developing chemical links between a number of monomers or repeating units. For instance, polymerizing styrene, a monomer, under suitable reaction conditions, results in the polymer polystyrene.

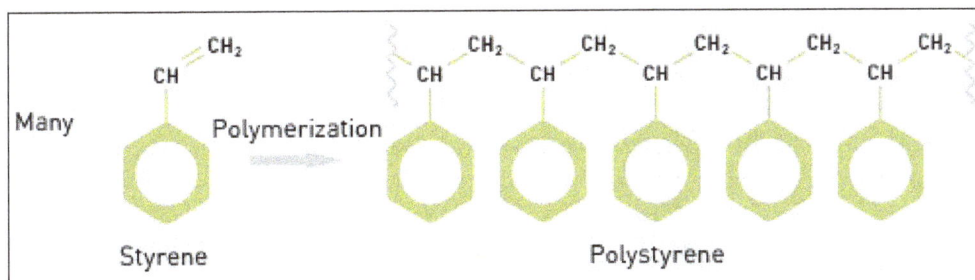

A defining feature of polymers is their chain-like structure, made up of repeating monomers.

A polymer's molecular weight is related to that of the monomer, and the number of monomers present in the polymer molecule. The molecular weight of styrene is 104 Da. Hence, the molecular weight of polystyrene is 104n, where "n" is the styrene molecule number in the polymer chain. Both the distribution shape and the average MW influence the properties of a polymer. Therefore, measuring MW requires measuring the MW of individual chains and the number of chains of any specific weight.

The general distribution of polymer MW is seen in figure. Using statistics, three different moments can be defined for this distribution, each being considered as an average MW. Mn is the number averaged MW, and Mw is the weight averaged MW. The midpoint of the distribution in terms of the number of molecules is Mw. The third moment, Mz, has more weighting with regards to higher MWs. The Mw:Mn ratio is termed as polydispersity, and is used for describing the distribution width.

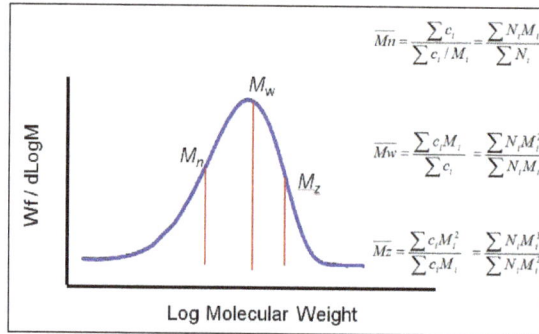

$$\overline{Mn} = \frac{\sum c_i}{\sum c_i / M_i} = \frac{\sum N_i M_i}{\sum N_i}$$

$$\overline{Mw} = \frac{\sum c_i M_i}{\sum c_i} = \frac{\sum N_i M_i^2}{\sum N_i M_i}$$

$$\overline{Mz} = \frac{\sum c_i M_i^2}{\sum c_i M_i} = \frac{\sum N_i M_i^3}{\sum N_i M_i^2}$$

Average MW can be defined in a number of different ways using different moments of the distribution.

An essential point to consider while examining alternating MW technique measurements is that any absolute MW measurement must involve the absolute measurement of concentration or number of molecules.

Measuring MW - Introducing Static Light Scattering (SLS)

The different methods used for measuring MW are:

- End group analysis

- Membrane osmometry

- Viscometry

- Light scattering

The standard approach, however, is static light scattering (SLS). The irradiation of a macromolecule by an incident light photon beam causes photons to be absorbed and re-emitted or scattered in all directions. The scattered light intensity and the polymer MW are proportional, the relationship being described by the Rayleigh equation.

$$\frac{KC}{R_\theta} = \left(\frac{1}{Mw} + 2A_2 C \right) \frac{1}{P_\theta}$$

Three approaches to SLS have been developed:

- Right angle light scattering (RALS) - Scattering intensity is determined at 90° to the incident beam. It offers a very good signal to noise ratio, however, does not take into account anisotropic scattering. The assumption is that the scattering intensity at 0° is just as that at 90°. This is a suitable approach for small molecules but not for anisotropic scatterers.

- Low angle light scattering (LALS) - In LALS, scattering intensity is determined at an angle very close to 0° in order to eliminate the error related to anisotropic scattering. This is fine for all molecules, however, the signal-to-noise ratio becomes challenging for smaller molecules. A combination of RALS/LALS technology is a good prospect.

- Multi-angle light scattering (MALS) – The approach adopted with MALS is measuring at several angles and extrapolating the same to determine a value for scattering intensity at 0 °C. MALS is suitable for all molecule sizes but the method is more complicated than LALS or RAls.

It is possible to use all the three light scattering techniques in batch mode, but they are more frequently applied in flow mode, with the light scattering detector constituting a gel permeation/size exclusion chromatography (GPC/SEC) detector array.

GPC/SEC - Powerful Technique for Measuring MW Distribution

As shown in figure, GPC/SEC starts with size fractionation of a sample, after which each sample fraction is detected as it elutes from the separation column. The only disadvantage with GPC/SEC is that separation is based on the polymer molecule size, not on its MW.

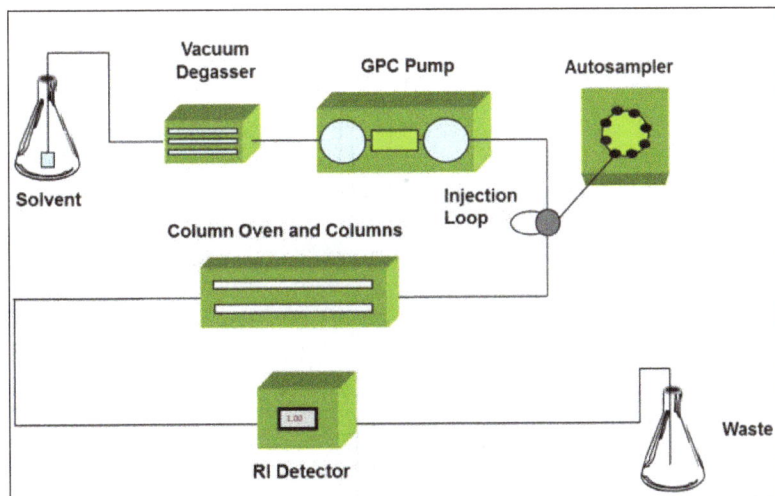

Schematic of a GPC/SEC set-up, detectors are positioned at the exit of the column oven to measure the properties of the eluting sample.

In GPC/SEC, the term 'absolute' MW is used for differentiating a method that directly measures MW from one that infers it, especially from calibration techniques relying on the use of a set of relevant standards.

Light scattering detectors are a convenient tool for determining MW in a GPC/SEC system set-up. SLS provides the most precise data available with present technology, and can be considered absolute in the sense of direct measurement.

Importance of Molecular Size

Molecular size, similarly to MW, is a defining property of most polymers. A polymer molecule's size in solution directly impacts its rheological behavior, which is linked directly to formulation performance.

MW and molecular size relationship is not constant, and different polymers show different molecular densities in solution.

Most polymer molecules are below 100 nm, and hence, the requirement for size information calls for techniques that have the ability to measure nanoparticles. Three different techniques are applied: dynamic light scattering (DLS); MALS; and SLS in combination with intrinsic viscosity (IV) measurement.

Measuring Molecular Size (1) DLS

When compared to SLS, DLS measurements derive from the real-time fluctuations in scattered light, instead of the time-averaged data, which is the focus in the case of static techniques. DLS is mostly applied in batch mode. The illumination of a sample is performed with a laser beam, resulting in light scattering that is detected by a sensitive photon counting module.

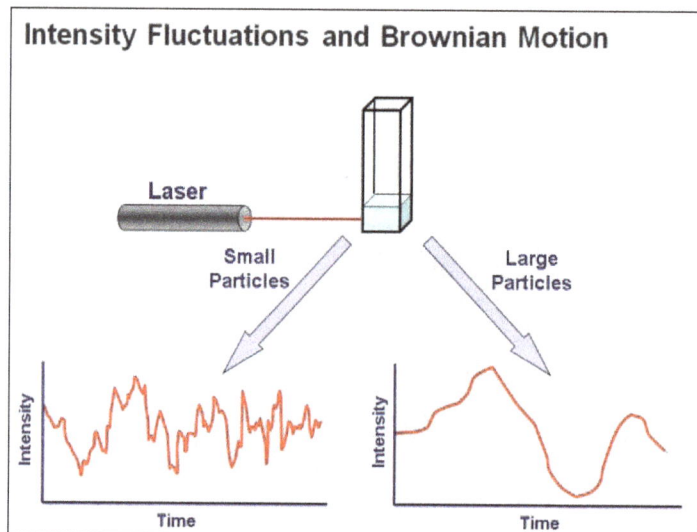

DLS measurements determine particle size from the pattern of intensity fluctuations in scattered light.

A correlator is used to translate the determined light scattering intensity fluctuations to a measure of diffusion speed, thereby providing a value for molecule or particle size. The Stokes-Einstein equation clearly explains the relationship between size and speed of motion. Determining the scattered light intensity helps to determine particle or molecular size, more specifically hydrodynamic radius (Rh).

Measuring Molecular Size (2) MALS

MALS is used for determining the molecular size based on the fact that the anisotropic light scattering pattern produced by larger molecules of around 10 to 15 nm radius and above is associated with their size. For this reason, MALS cannot determine molecular size for smaller molecules. For larger molecules, the radius of gyration (Rg) can be computed.

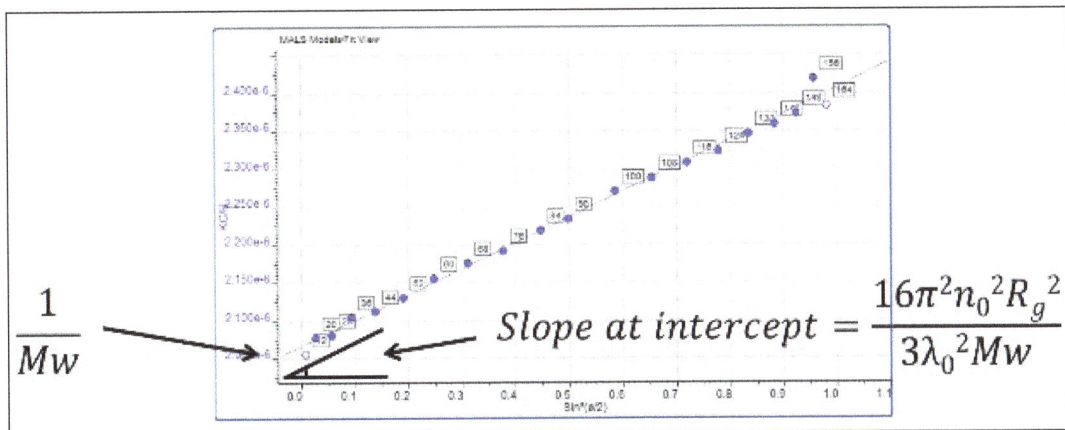

$$\frac{1}{Mw} \qquad Slope\ at\ intercept = \frac{16\pi^2 n_0^2 R_g^2}{3\lambda_0^2 Mw}$$

Rg the radius of gyration can be determined from a Zimm plot which can be produced by measuring light scattering intensity as a function of angle for particles that scatter anisotropically (> 10 - 15 nm in size).

Measuring Molecular Size (3) - SLS with IV

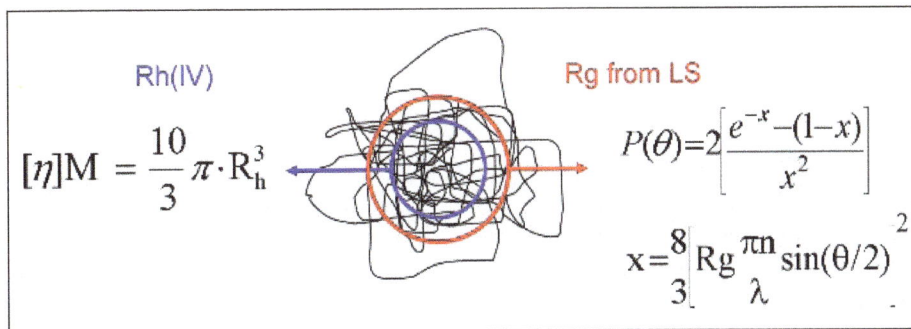

Rh(IV)   Rg from LS

$$[\eta]M = \frac{10}{3}\pi \cdot R_h^3$$

$$P(\theta)=2\left[\frac{e^{-x}-(1-x)}{x^2}\right]$$

$$x=\frac{8}{3}\left[Rg\frac{\pi n}{\lambda}\sin(\theta/2)\right]^2$$

Rh (IV) is a different size metric to Rg measured by MALS.

Even though the relation between a polymer's molecular size and the MW is not constant, it can be determined. The parameter IV, measured in units dl/g, can be determined with viscometry measurements, and relates molecular size directly to MW for any particular polymer. IV is an inverse measure of molecular density. Polymers that are less tightly coiled have a considerably high IV, while the densely entangled ones have a low IV. Molecular size can be determined from MW measurements by SLS combined with IV. This method helps obtain values for hydrodynamic radius (Rh). It is important to note that Rh is different from Rg.

Comparing Measures of Polymer Molecular Size

Different size parameters are obtained from the three methods detailed for molecular size measurement, Rh (DLS), Rg and Rh(IV). The magnitude of any difference in these three parameters is based on the molecular density and shape of that particular molecule. Molecular size, like MW, is a distributed parameter.

Investigating Molecular Structure the Link between Size and MW

It is possible to determine molecular size and MW distributions for a polymer using GPC/SEC and the aforementioned measurement methods. The explanation of structural characteristics depends on the use of these data along with IV measurements, which is a measure of molecular density. Table shows the main ways in which IV is changed by fluctuations in the structure, relationships which underpin the value of adding a viscometer to a GPC/SEC system to determine IV, and which hence explain structure.

Table: Changes in polymer structure have a direct impact on IV which is consequently a useful metric for investigating structural characteristics.

Structural or conformational change	Effect on density	Effect on IV
a) Increase chain length (MW) of linear molecule.	Decreases	Increases according to Mark-Houwink equation
b) Increase mass of chain segments, keeping chain length constant.	Increases	Decreases
c) Increase stiffness of chain.	Decreases	Increases
d) Add branches to chain, keeping MW constant.	Increases	Decreases
e) Collapse chain into dense molecule. (natural protein or aggregate)	Increases greatly	Decreases greatly

IV can be determined in continuous mode or by batch techniques in a GPC/SEC set-up, but in both cases, the relative or specific viscosity is measured. Figure shows a viscometer design specified for use in a GPC/SEC setup, which works similarly to a Wheatstone bridge on the balancing resistances principle.

Differential pressure transducers determine the pressure drop across the center of the bridge, DP and from inlet to outlet, IP. The main design features are that the delay volume must be higher than the net elution volume of the separation column, however, with negligible flow resistance.

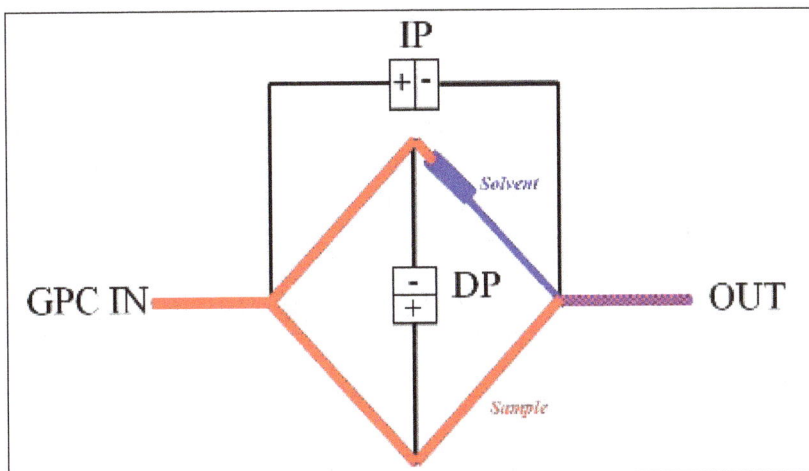

Schematic of differential viscometer which can be used within a GPC/SEC set-up to determine IV.

Using differences in IV, data about the structure using correlations outlined in table, detailed data about structural characteristics can be determined from a Mark-Houwink (M-H) plot and a log-log plot of IV against MW.

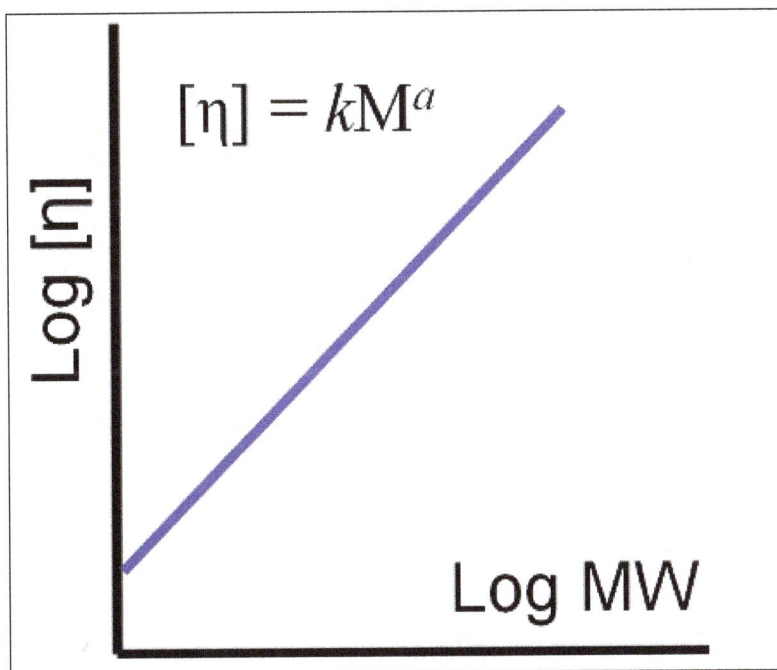

$$[\eta] = kM^a$$

The shape of the M-H plot, gradient and intercept, enables the robust comparison of different polymer samples in terms of their structure.

The M-H parameters, k and a, as shown in figure, can be seen from the M-H plot. Figure shows an M-H plot for a considerably high MW, standard linear polystyrene in THF. This was obtained in a SEC/GPC experiment by accurately determining the MW (light scattering) and IV (viscometer) of the eluting sample. The result is a precise straight line M-H plot from which k and a values can be easily determined. The straight line shows that the polymer structure in solution does not change across all the molecular weights it contains.

The same data for a sample containing some low MW polystyrene in addition to the original material are shown in figure. Plots for three Maltodextrin batches are shown in figure, the properties of which are shown in the inset table. The bulk IV values of these samples are very similar.

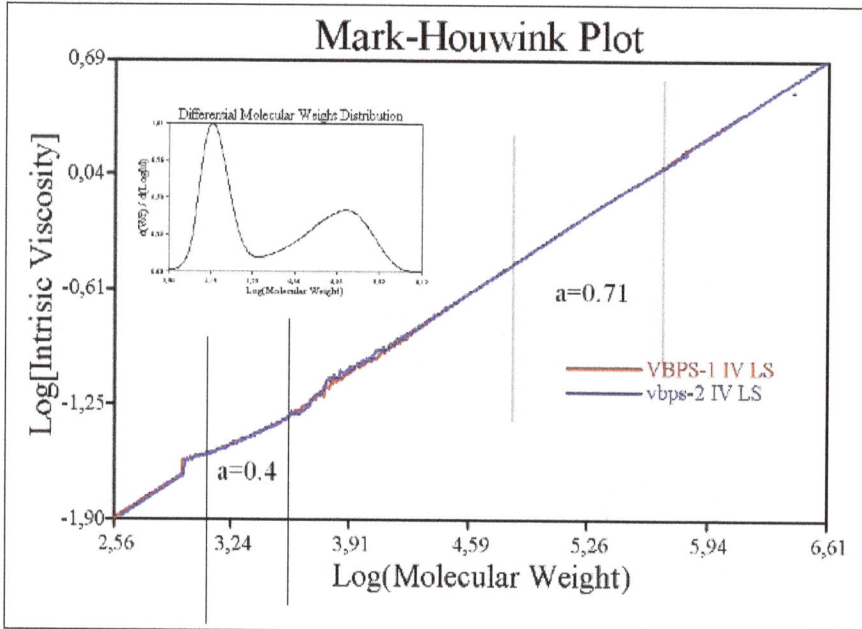

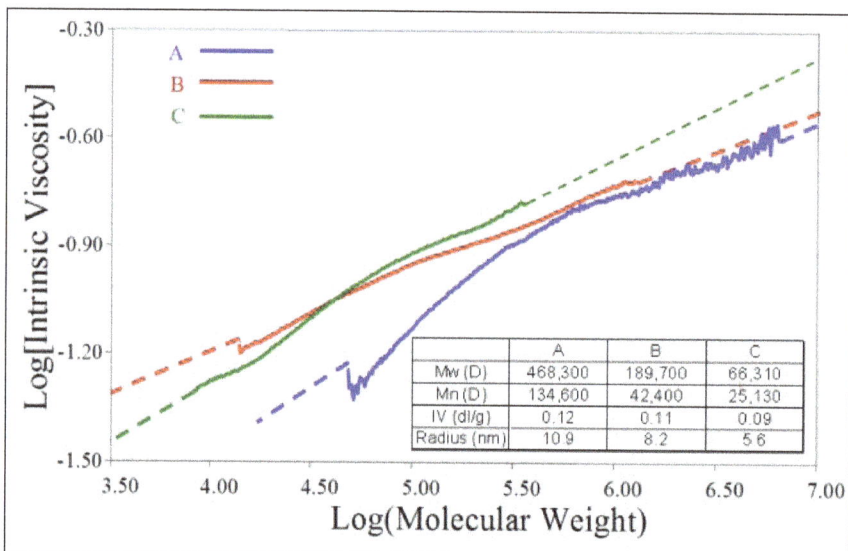

M-H plots for three maltodextrin samples with closely similar IV values reveal marked differences in structure.

References

- Polymer, science: britannica.com, Retrieved 28 April, 2019
- Polymer-chemistry, earth-and-planetary-sciences: sciencedirect.com, Retrieved 16 June, 2019
- Classification-of-polymers, chemistry-polymers: toppr.com, Retrieved 08 July, 2019
- Polymers: byjus.com, Retrieved 05 August, 2019

2
Organic Polymers

The polymers which contain carbon atoms are known as organic polymers. Some of the various types of organic polymers are polyvinyl ether, polypyrrole, poly(diiododiacetylene), polydioctylfluorene, polyorthoester, polyphenylene sulphide and polythiophene. This chapter has been carefully written to provide an easy understanding of these types of organic polymers.

Organic polymers are polymer materials that essentially contain carbon atoms in the backbone. Therefore, there are only carbon-carbon covalent bonds in these. These polymers form only from organic monomer molecules. Most of the times, these polymers are environmentally friendly since these are biodegradable.

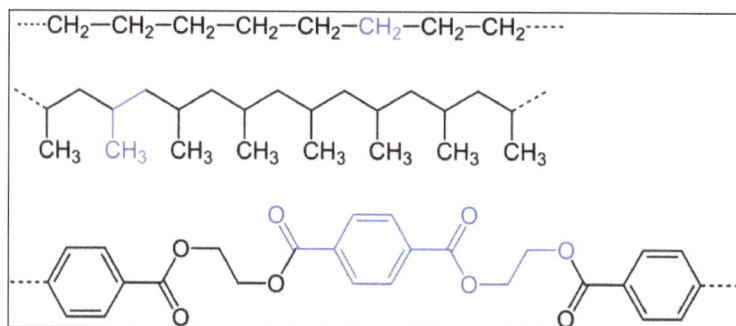

Some Examples of Organic Polymers.

Furthermore, there are two major forms of organic polymers such as natural and synthetic polymers. Common examples of important organic polymers include polysaccharides, proteins, polynucleotides (DNA and RNA), etc. These are natural organic polymers. Synthetic organic polymers include polyesters, nylon, polycarbonate, etc.

Similarities between Organic and Inorganic Polymers

- Both are polymer materials consisting of monomers that are linked to each other via covalent bonds.

- Both Organic and Inorganic Polymers are macromolecules having very high molar masses.

Difference between Organic and Inorganic Polymers

Organic polymers are polymer materials that essentially contain carbon atoms in the backbone. These polymers essentially contain carbon atoms in the backbone. Most of the organic polymers are simple structures. Moreover, these are environmentally friendly since these are biodegradable. On the other hand, inorganic polymers are polymer materials that do not have carbon atoms in the backbone. Therefore, these polymers do not contain carbon atoms in the backbone. This is the main difference between organic nd inorganic polymers. Almost all of the inorganic polymers are highly branched complex structures. In addition, these are not environmental friendly because these are non-biodegradable.

Polyvinyl Ether

Polyvinyl ethers are either viscous oils or soft and tacky rubbery materials depending on the structure and molecular weight. Their glass transition temperature is well below room temperature. They are known for their elasticity and excellent resistance to hydrolysis (being ether-based, PVEs do not hydrolyze). They also have fairly good weatherability and heat resistance since they do not have double bonds in the backbone. They are miscible with water and are soluble in many solvents. They are usually none hazardous substances and not irritants to skin.

Vinylether monomers have unique properties. Due to the presence of the neighboring oxygen atom, the double bond is highly electronegative. For this reason, they easily undergo (living / controlled) cationic polymerization. This makes them very versatile curing agents. The reactivity of the vinyl ethers in cationic polymerization depends on both the initiator used and the structure of the vinyl ether itself. In general, vinyl ethers possessing highly branched alkyl groups are more reactive than those bearing straight-chain alkyl groups.

Free-radical copolymerization of vinyl ethers is also possible; it can be initiated with peroxide, azo, and redox initiators. However, the polymerization under free radical conditions gives only low-molecular-weight oligomers and only copolymers can be synthesized. Furthermore, in the presence of water they readily hydrolyse to acetaldehyde and alcohol below pH of about 5.5 which makes emulsion polymerization with certain monomers like vinyl acetate difficult unless the pH is carefully controlled.

Some commercially important vinyl ethers include ethyl, n-butyl, isobutyl, ethylhexyl, dodecyl, octadecyl, and cyclohexyl vinylether.

Applications

Poly(vinyl ethers) and their copolymers find many applications such as adhesives, release and surface coatings, lubricants, greases, elastomers, anticorrosion agents, molding compounds, personal care products, fiber and textile finishes.

Polyvinyl ethers are widely used in adhesive and coating formulations. Due to their efficiency as reactive diluents and their ability to undergo both cationic homopolymerization and free-radical copolymerization, vinylether monomers are excellent co-reactants for silicone release coatings and silicone pressure sensitive adhesives. Some benefits of the addition of vinylethers to commercial (UV curable) epoxysilicone release coatings include improved cationic photoinitiator miscibility, viscosity reduction, and lower raw material costs. Vinylether monomers are not inexpensive, however, they are less costly than epoxy silicone polymers and hence, lower raw material costs. Vinylethers are also known to be effective reactive diluents for acrylate oligomers in free radical and hybrid adhesives. Like acrylic polymers, they provide pressure sensitive adhesives with a high degree of tack, hence, no tackifier resins have to be added. Due to their high(er) cost, but excellent compabiltiy with acrylic resins, they are ususally blended with acrylic polymers/oligomers to improve certain characteristics of the PSA. Due to their good adhesion to skin and high water permeability, they are sometimes used for medical pressure sensitive tapes and dressings.

Polyvinyl ethers are also used as synthetic lubricants, mainly as refrigerant oils. The synthetic vinyl ether oils (PVE) have excellent lubricating properties. However, they provide not better lubrication than petroleum-based lubricants such as mineral oils, but they do have a narrow molecular weight distribution which gives the oil very consistent physical properties. Furthermore, the polymer-based chemical structure and controllable synthesis makes it possible to tailor the PVE oil with different viscosity and miscibility properties. Compared with polyolester oils (POE), they are more chemically stable since carbon-oxygen double bonds are absent.

Polypyrrole

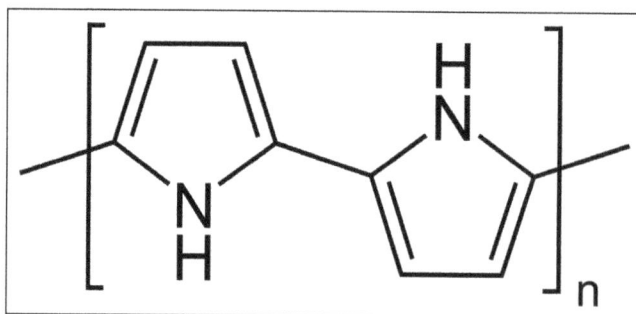

Polypyrrole.

Polypyrrole (PPy) is a type of organic polymer formed by the polymerization of pyrrole.

It is a solid with the formula $H(C_4H_2NH)_nH$. Upon oxidation, polypyrrole converts to a conducting polymer.

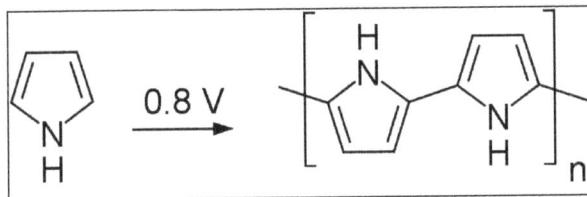

Pyrrole can be polymerised electrochemically.

Synthesis

Polypyrrole is prepared by oxidation of pyrrole:

$$n\ C_4H_2NH + 2n\ FeCl_3 \rightarrow (C_4H_2NH)_n + 2n\ FeCl_2 + 2n\ HCl$$

The process is thought to occur via the formation of the pi-radical cation $C_4H_4NH^+$. This electrophile attacks the C-2 carbon of an unoxidized molecule of pyrrole to give a dimeric cation $(C_4H_4NH)_2]^{++}$. The process repeats itself many times.

Conductive forms of PPy are prepared by oxidation ("p-doping") of the polymer:

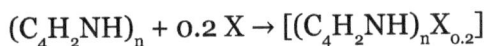

$$(C_4H_2NH)_n + 0.2\ X \rightarrow [(C_4H_2NH)_nX_{0.2}]$$

The polymerization and p-doping can also be effected electrochemically. The resulting conductive polymer are peeled off of the anode. Cyclic voltammetry and chronocoulometry methods can be used for electrochemical synthesis of polypyrrole.

Properties

Films of PPy are yellow but darken in air due to some oxidation. Doped films are blue or black depending on the degree of polymerization and film thickness. They are amorphous, showing only weak diffraction. PPy is described as "quasi-unidimensional" vs one-dimensional since there is some crosslinking and chain hopping. Undoped and doped films are insoluble in solvents but swellable. Doping makes the materials brittle. They are stable in air up to 150 °C at which temperature the dopant starts to evolve (e.g., as HCl).

PPy is an insulator, but its oxidized derivatives are good electrical conductors. The conductivity of the material depends on the conditions and reagents used in the oxidation. Conductivities range from 2 to 100 S/cm. Higher conductivities are associated with larger anions, such as tosylate. Doping the polymer requires that the material swell to accommodate the charge-compensating anions. The physical changes associated with this charging and discharging has been discussed as a form of artificial muscle. The surface of polypyrrole films present fractal properties and ionic diffusion through them show anomalous diffusion pattern.

Applications

PPy and related conductive polymers have two main application in electronic devices and for chemical sensors.

Research Trends

PPy is a potential vehicle for drug delivery. The polymer matrix serves as a container for proteins.

Polypyrrole has been investigated as a catalyst support for fuel cells and to sensitize cathode electrocatalysts.

Together with other conjugated polymers such as polyaniline, poly(ethylenedioxythiophene) etc., polypyrrole has been studied as a material for "artificial muscles", a technology that offers advantages relative to traditional motor actuating elements.

Polypyrrole was used to coat silica and reverse phase silica to yield a material capable of anion exchange and exhibiting hydrophobic interactions.

Polypyrrole was used in the microwave fabrication of multiwalled carbon nanotubes, a rapid method to grow CNT's.

A water-resistant polyurethane sponge coated with a thin layer of polypyrrole absorbs 20 times its weight in oil and is reusable.

The wet-spun polypyrrole fibre can be prepared chemical polymerization pyrrole and DEHS as dopant.

Poly(diiododiacetylene)

PIDA, or poly(diiododiacetylene), is an organic polymer that has a polydiacetylene backbone. It is one of the simplest polydiacetylenes that has been synthesized, having only iodine atoms as side chains. It is created by 1,4 topochemical polymerization of diiodobutadiyne. It has many implications in the field of polymer chemistry as it can be viewed as a precursor to other polydiacetylenes by replacing iodine atoms with other side chains using organic synthesis, or as an iodinated form of the carbon allotrope carbyne.

Structure

The backbone of PIDA is highly conjugated and allows for the formation of an extended pi system along the length of the polymer. This property of PIDA allows it to transport electricity and act as a molecular wire or an organic semiconductor. Considering

PIDA's backbone and the fact that Iodine atoms can easily undergo elimination, it is conceivable that PIDA can be subjected to full reductive deiodination in the presence of a Lewis base, such as pyrrolidine to yield carbyne.

Synthesis

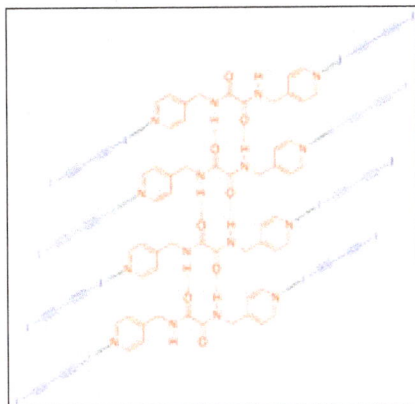

Structure of co-crystals. The oxalamide host is shown in red and diiodobutadiyne in blue.

PIDA is synthesized from diiodobutadiyne via 1,4 topochemical polymerization. In order to meet the geometric requirements for polymerization, a host–guest strategy is used by combining a host molecule and diiodobutadiyne in solution and allowing co-crystallization to occur. This can be utilized because hosts that are most commonly used are able to bond to the diyne monomer by halogen bonding from the lewis acidic iodine atom to a lewis basic nitrogen of the host (usually a nitrile or pyridine). In order to give a proper repeat distance to the monomers (5 Å), the hosts also contain oxalamide groups that create a hydrogen bonding network throughout the crystal.

In most instances, polymerization is spontaneous upon crystallization or exposure to UV radiation/pressure.

Reactions

PIDA Can undergo carbonization at high temperatures near 900 °C and reductive dehalogenation carbonization when exposed to pyrrolidine at room temperature.

Attempts have been made to replace iodine side groups with other functional groups. There are also attempts being made at making other halogen analogs of PIDA.

Polydioctylfluorene

Polydioctylfluorene (PFO) is an organic compound, a polymer of 9,9-dioctylfluorene, with formula $(C_{13}H_6(C_8H_{17})_2)_n$. It is an electroluminescent conductive polymer that characteristically emits blue light. Like other polyfluorene polymers, it has been studied as a possible material for light-emitting diodes.

Structure

The monomer has an aromatic fluorene core $-C_{13}H_6-$ with two aliphatic n-octyl $-C_8H_{17}$ tails attached to the central carbon. Polydioctylfluorene (PFO) can be found in liquid-crystalline, glassy, amorphous, semi-crystalline or β-chain formation. This variety is on account of the intermolecular forces that PFO can participate in. The secondary forces present in PFO are typically van der Waals, which are relatively weak. These weak forces makes it a solid that can also be used as a film on a substrate. The glassy films formed by PFO chains form solutions in good solvents, meaning it is at least partially soluble. These van der Waals also add complexity to the microstructure of PFO, which is why it has a wide range of solid formations. The solid formations though, typically form low density due to the low cooling rate of the polymer. The density of polydioctylfluorene is measured by using the process of ultraviolet photoelectron spectroscopy. Chain stiffness is also prominent in PFO, because of this it is predicted that the molecular weight is a factor of 2.7 lower than polystyrene, which can produce an approximation of 190 repeat units in a standard PFO chain. By changing the strain and temperature applied to the polymer's structure results in an alteration of the PFO's properties. Thermal treatment such as friction transfer can be applied to the structure, this is a way to alter the properties. The friction transfer aligns the structure to become crystalline or liquid crystalline. Polymer 196 is the most commonly studied type of polydioctylfluorene. In studies, polymer 196 has shown the most promising properties and the best crystallinity. Within the crystal structure of polymer 196 octyl side chains are inserted between the layer of the polymer to provide more space for efficiency in structuring the material.

In studies, the structure of polydioctylfluorene was observed by using grazing-incidence X-ray diffraction after applying friction to the structure. Experiments revealed PFO was present in crystalline films and liquid crystalline after cooling and use of friction. As a result of the friction exerted, twofold symmetry in PFO was broken. The friction transfer used to obtain a single crystal film is important in the process of fabricating polarized light emitting diodes.

Properties

Polydioctylfluorene, can also be known as polymer 196 to polyfluorene. The molar mass of PFO ranges between 24,000-41,600 (g/mol) and because of this varying molar mass, many other properties vary as well. For example, the glass transition temperature can fall somewhere between 72-113 degrees Celsius. The absolute wavelength emitted by PFO can range between 386-389 nm in a solution of $CHCl_3$, and falls around 389 in a solution of THF. The absolute film wavelength of PFO though falls between 380-394 nm. The melting point of a crystalline molecule of PFO is predicted to be about 150 degrees Celsius.

There have also been reports that some of the solid states of polydioctylfluorene are composted in sheet-like layers which are about 50-100 nm thick. As a result of these sheets, the glassy and semicrystalline states can be formed (excluding amorphous, liquid crystalline, and beta chain states). When cooled quickly, the chains tightly align, giving PFO a close packing factor, though because of the high complexity of the chains, this sometimes gets messy and creates the amorphous state. The parts of the molecule that add this complexity are the carbon rings (that are located in the backbone) making the molecule overall large in size.

Applications

The formation of beta-phase chains in PFO can be formed through dip-pen nanolithography, to represent wavelength changes in metamaterials. The dip-pen technique allows a scale of 500 nm > to be visible. The beta chains can be converted into the glassy films by adding extra stress to the main fluorine backbone unit, whether beta chains are formed is determined by peaks in wavelength absorption. Beta chains can also be confirmed to be present by using solvent to non-solvent mixtures. If the molecule were to be dipped into this mixture for ten seconds, the chains with no dissolution of films are able to produce these said beta chains.

Polydioctylfluorene is a polymer light-emitting device known as PLED, which covalently bonds to the carbon hydrogen chains. PFO is a copolymer of basic polyfluorene, which enables it to release phosphorescent light. This basic fluorene backbone strengthens the molecule on account of the carbon rings. The cross-linking in polydioctylfluorene structure provides an efficient technique for hole-transport layers to emit light. Also, when a solvent-polymer compound is added the β-phase crystalline structure to be maintained. Efficiency in current can reach a maximum of about 17 cd/A and maximum luminance obtained can be approximately 14,000 cd/m(2). The hole-transport layers (HTLs) improve the polymer's anode hole injection and greatly increase electron blocking. By having the capability to control the microstructure of phase domains gives an opportunity to optimize the optoelectronic properties of PFO based products. When needs for optoelectronic emittance are reached in polydioctylfluorene, the electroluminescence given off in dependent on the active layer in the conjugate polymer. Another way to affect the optoelectronic properties is by altering how dense the phase chain segments are ordered. Low densities can be achieved from tremendously slow crystallization while on the other hand directional crystalline solution can be achieved by use of thermal gradients.

Polyorthoester

Polyorthoesters are polymers with the general structure $-[-R-O-C(R_1, OR_2)-O-R_3-]_n-$ whereas the residue R_2 can also be part of a heterocyclic ring with the residue R.

Polyorthoesters are formed by transesterification of orthoesters with diols or by polyad-dition between a diol and a diketene acetal, such as 3,9-diethylidene-2,4,8,10-tetraox-aspiro undecane.

Applications

Polyorthoesters are used as hydrophobic implant materials for drug depots for contin-uous drug delivery by surface erosion. The active ingredient (which is homogeneously dispersed in a matrix of polyorthoester) should be released as evenly as possible into the human or animal organism over an extended period of time in a zero-order release kinet-ics. Four classes of polyorthoesters (polyorthoesters type I - IV) are well characterized as biodegradable polymers for drug implants, primarily through work of Jorge Heller.

Production

1st Generation Polyorthoester: POE I

Polyorthoester type I is (usually) obtained by transesterification of an α,ω-di-ol with 2,2-diethoxytetrahydrofuran (synthesized from γ-butyrolactone and triethylorthoformate).

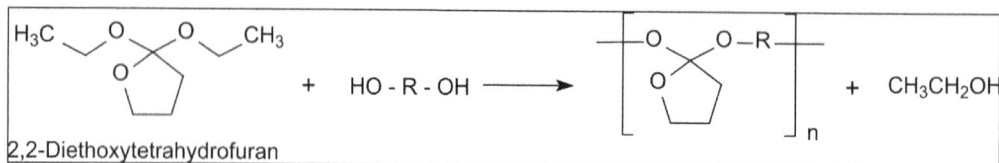

2,2-Diethoxytetrahydrofuran

In polycondensation small molecules are formed (in this case of ethanol), which have to be removed from the equilibrium to achieve the necessary molar mass of the polymer for the use as an implant material. The solid polyorthoester type I is hydrophobic and particularly acid-sensitive. In an aquatic environment it autocatalytically hydrolysis in an uncontrolled fashion. Therefore, it must be stabilized by adding an alkaline pharma-ceutical excipient when used as an implant material.

The degradation of the polymer chain sets free the initial diol and γ-butyrolactone, which is further hydrolyzed to 4-hydroxybutanoic acid. The 4-hydroxybutanoic acid formed is responsible for the locally lowered pH value upon polymer degradation.

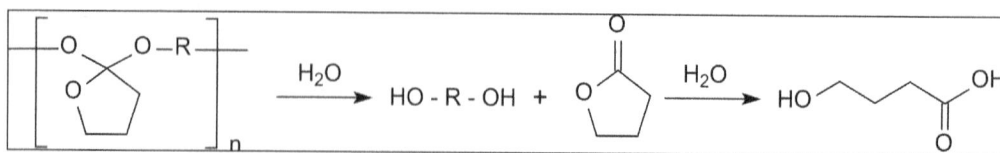

The commercial use of polyorthoester type I was prevented by the required addition of a base (e.g. sodium carbonate), the difficult synthesis and its unsatisfactory mechanical properties.

2nd Generation Polyorthoesters: POE II

Polyorthoesters type II are formed by polyaddition of a α,ω-diol and the diketene acetal 3,9-diethylidene-2,4,8,10-tetraoxaspiro undecane (DETOSU). The polyaddition forms much more quickly high molecular weight polymers than the transesterification does and in contrast to polyorthoester type I no small molecules are released. For the reaction, the monomers are dissolved in tetrahydrofuran and small amounts of an acidic catalyst are added, e. g. p-toluenesulfonic acid. The molecular weight of the polymers can be controlled by the molar ratio of the reactants. The addition of triols leads to crosslinked polymers, whereas the crosslinking density is determined by the ratio of triol/diol. The polymerization takes already place rapidly at room temperature and ambient pressure and allows the formation of a polymer matrix in the presence of sensitive pharmaceutically active agents.

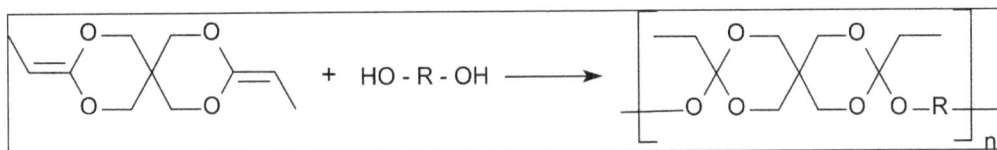

The solid polyorthoester type II polymers are very hydrophobic, storable in the dry and significantly less sensitive to acid than polyorthoester type I. The pH-sensitivity (and thus the rate of degradation in physiological media) as well as the glass transition temperature (and thus the mechanical and thermal properties) can be controlled through the use of diols of different chain flexibility. polyorthoester type II with molecular weights of up to about 100,000 have therefore a glassy-hard (e. g. when using the rigid 1,4-cyclohexanedimethanol) to semi-soft consistency (when using the flexible 1,6-hexanediol). In the aqueous medium a two-stage, non-autocatalytic hydrolysis takes place, initially generating neutral fragments (pentaerythritol dipropionate and the diol).

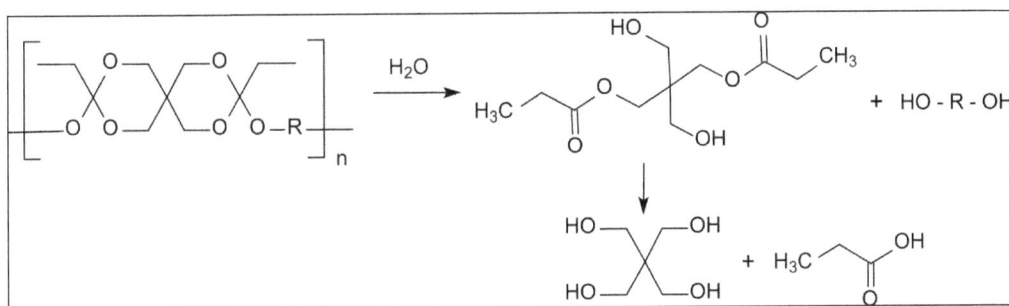

The propionic acid produced in the second step is metabolized so rapidly that a local lowering of the pH value does occur. Therefore, to accelerate polymer degradation acidic additives must be added (such as octanedioic acid, hexanedioic acid or 2-methylidenebutanedioic acid). Zero-order release kinetics were achieved when embedding the cytostatic agent 5-fluorouracil. In toxicity tests as specified in the US Pharmacopeia USP polyorthoester preparations were found to be acutely nontoxic in cellular, intradermal, systemic and intramuscular implants.

3rd Generation Polyorthoester: POE III

Polyorthoester type III is prepared just like POE I by transesterification, in this case a triol (preferably 1,2,6-hexanetriol) with an orthoester (e. g. triethylorthoacetate).

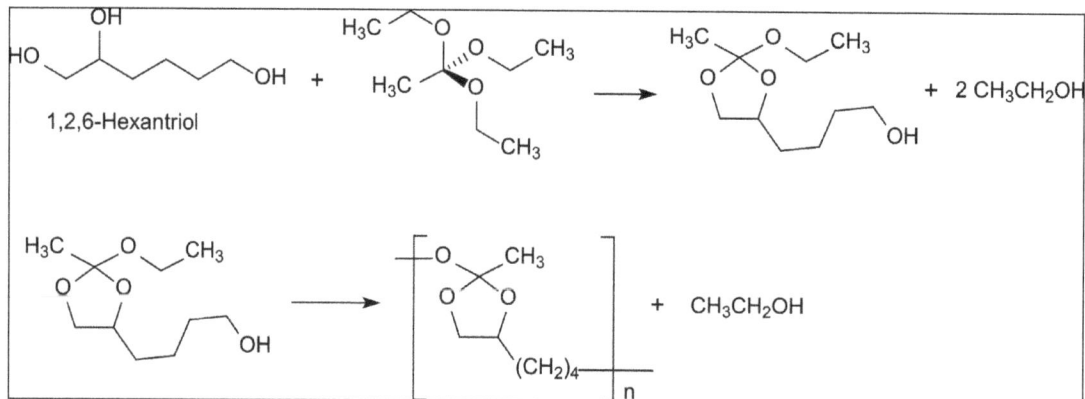

The triethylorthoacetate reacts initially to the corresponding cyclic orthoester with the vicinal hydroxyl groups of the 1,2,6-hexanetriol, which is homopolymerized to polyorthoester type III by reaction with the 6-position hydroxyl group. Polyorthoesters type III are at room temperature semi-solid to ointment-like due to the very flexible polymer backbone. They allow the incorporation of thermally labile and solvent-sensitive active ingredients at room temperature without the use of organic solvents. Such drug implants are particularly suitable for applications on the eye, where no sudden release occurs by diffusion (initial burst release) but the release follows the continuous polymer degradation. Also for Polyorthoesters type III the degradation occurs at the surface by cleavage of the hydrolytically labile bonds in the polymer backbone.

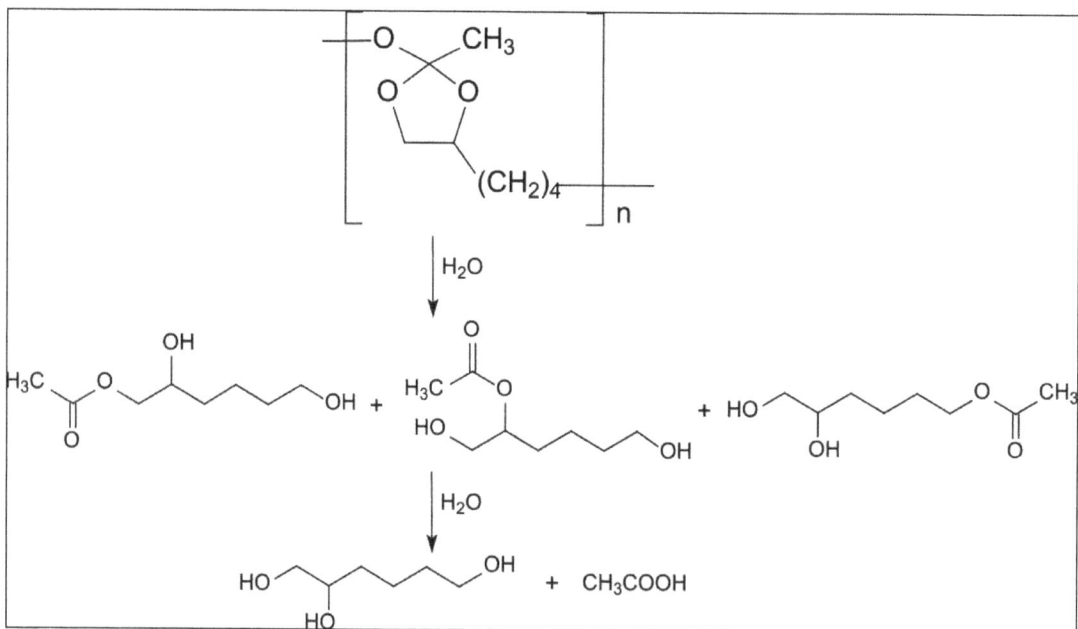

Depending on the initial bond cleavage on the quaternary carbon atom 1-, 2-, or 6-acetoxy-hexanetriol is formed, which is further degraded to acetic acid and 1,2,6-hexanetriol. The use of polyorthoester type III for biomedical applications is severely limited by the lengthy synthesis of polymers having useful molecular weights and poor reproducibility.

4th Generation Polyorthoesters: POE IV

The polyorthoester type IV is a further development of the type polyorthoester type II, which is formed of the diketene acetal DETOSU with a diol which is modified by short sequences of polyglycolide or polylactide. Depending on the type of diol used polyorthoester type IV can be synthesized as gel (with a low glass transition temperature Tg, meaning low molecular weight) or as a solid. Polyorthoester type IV-types are also accessible under the very mild conditions of interfacial polycondensation.

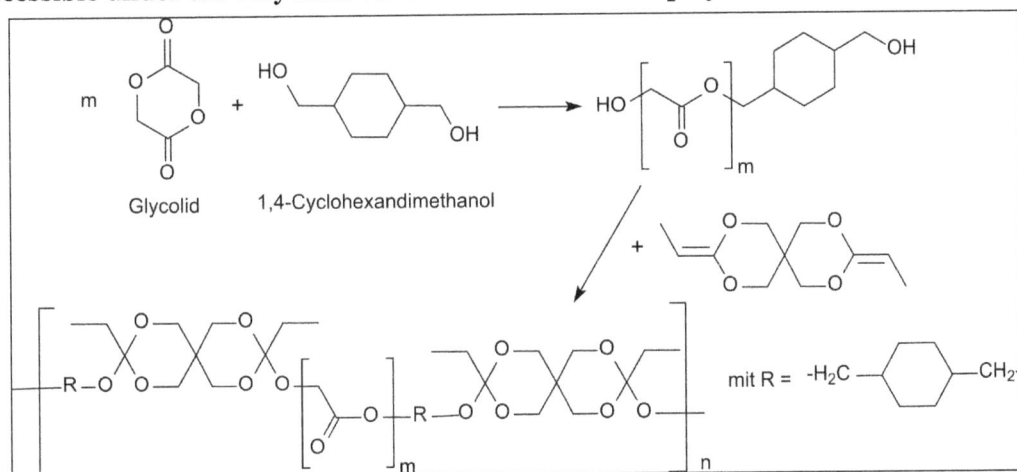

Glycolid   1,4-Cyclohexandimethanol

Polyorthoester type IV avoids the addition of acidic excipients required in polyorthoester type II, which often diffuse uncontrolled out of the polymer matrix and thus lead to erratic degradation kinetics. During the degradation of the polyorthoesters polyorthoester type IV in aqueous media glycolic acid or lactic acid is produced, which further catalyze hydrolysis.

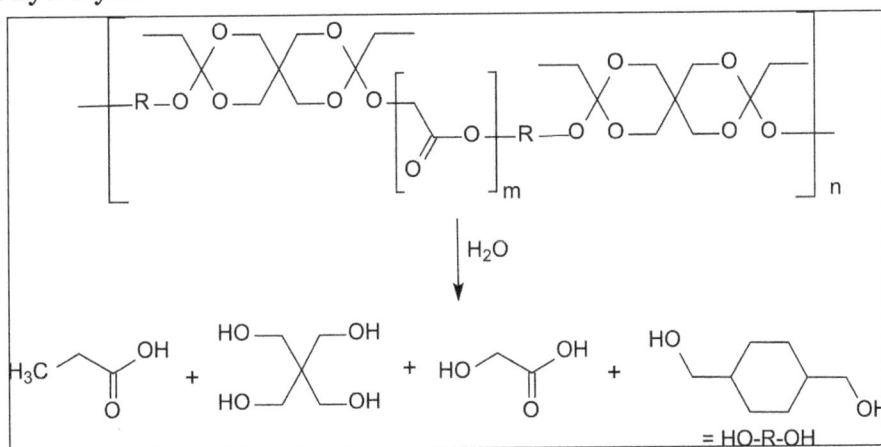

The degradation rate can be controlled by the proportion of glycolic or lactic acid in the sequence. Implants made of polyorthoester type IV show surface erosion while being highly biocompatible with degradation times from days to months and can thus also be used as a long-term drug depots, e. g. for the cytostatic agent 5-fluorouracil. Polyorthoesters type IV are considered the most promising members of this class as implant materials for controlled drug release.

Polyphenylene Sulfide

Polyphenylene sulfide (PPS) is an organic polymer consisting of aromatic rings linked by sulfides. Synthetic fiber and textiles derived from this polymer resist chemical and thermal attack. PPS is used in filter fabric for coal boilers, papermaking felts, electrical insulation, film capacitors, specialty membranes, gaskets, and packings. PPS is the precursor to a conductive polymer of the semi-flexible rod polymer family. The PPS, which is otherwise insulating, can be converted to the semiconducting form by oxidation or use of dopants.

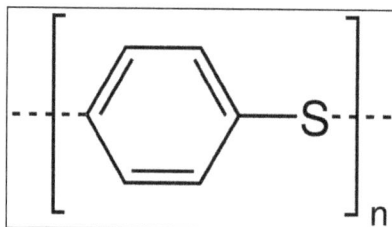
Polyphenylene sulfide.

Polyphenylene sulfide is an engineering plastic, commonly used today as a high-performance thermoplastic. PPS can be molded, extruded, or machined to tight tolerances. In its pure solid form, it may be opaque white to light tan in color. Maximum service temperature is 218 °C (424 °F). PPS has not been found to dissolve in any solvent at temperatures below approximately 200 °C (392 °F). An easy way to identify the compound is by the metallic sound it makes when struck.

Fiber Characteristics

PPS is one of the most important high temperature thermoplastic polymers because it exhibits a number of desirable properties. These properties include resistance to heat, acids, alkalies, mildew, bleaches, aging, sunlight, and abrasion. It absorbs only small amounts of solvents and resists dyeing.

Production

The Federal Trade Commission definition for sulfur fiber is "A manufactured fiber in which the fiber-forming substance is a long chain synthetic polysulfide in which at least 85% of the sulfide (—S—) linkages are attached directly to two (2) aromatic rings."

The PPS (polyphenylene sulfide) polymer is formed by reaction of sodium sulfide with p-dichlorobenzene:

$$ClC_6H_4Cl + Na_2S \rightarrow 1/n \, [C_6H_4S]_n + 2\,NaCl$$

The process for commercially producing PPS (Ryton) was initially developed by Dr. H. Wayne Hill Jr. and James T. Edmonds at Phillips Petroleum Company. NMP is used as the reaction solvent because it is stable at the high temperatures required for the synthesis and it dissolves both the sulfiding agent and the oligomeric intermediates.

Linear, high-molecular-weight PPS that is capable of being extruded into film and melt spun into fiber was invented by Robert W. Campbell.

The first U.S. commercial sulfur fiber was produced in 1983 by Phillips Fibers Corporation, a subsidiary of Phillips 66 Company.

Polythiophene

Polythiophene(PT) is a polymer which has drawn considerable attention in the field of third-order NLO properties. The band gap of approximately 2eV allows for investigations of NLO properties both under resonant conditions and off-resonance. The polymer has a very good stability in air and is easily doped. In addition to these advantages, PT can be made processible by attaching long flexible chains in the 3-position of thiophene. Thus, 3-substituted PTs are possible to study in several forms such as cast film, LB film or fiber. PT and its derivatives can be synthesized chemically or electrochemically. Some known 3-substituted PTs (65) are poly(3-hexylthiophene) (P3HT), poly(3-octylthiophene) (P3OT), poly(3-dodecylthiophene) (P3DDT), poly(3-octadecylthiophene) (P3ODT) and poly(3-eicosylthiophene) (P3ET).

Prasad *et al.* investigated the third-order NLO behavior of electrochemically polymerized PT using the DFWM method with 350fs pulses at 0.602 nm. At this wavelength the DFWM response was found to be governed by resonant nonlinearity. The effective value of $\chi^{(3)}$ was ~ 4×10^{-10} esu. A very interesting feature of the results described by Prasad *et al.* was the effect of doping on the third-order nonlinearity. Oxidizing (*p*-doping) of the PT film resulted in a drastic reduction of the effective $\chi^{(3)}$ value. Singh *et al.* investigated chemically and electrochemically polymerized films of P3DDT using the same technique. They found that the rise time of the nonlinear response was instantaneous and the decay involved a time constant of about 200fs. The measured effective $\chi^{(3)}$ values for the chemically and electrochemically prepared polymers were 3×10^{-10} esu and 5×10^{-10} esu respectively at 602 nm and 4.5×10^{-10} and 7×10^{-10} esu at 590 nm. Yang $et\ al.$ have reported $\chi^{(3)} = \sim 10^{-9}$ esu for unsubstituted PT around 600 nm. The observed value of $\chi^{(3)}$ at 705 nm was 4.0×10^{-11} esu for both samples.

Kobayashi and coworkers performed femtosecond transient absorption studies on PT and found decay times in the picosecond range. Similar results have been obtained by Vardeny $et\ al.$ who observed a complicated decay of transient absorption extending over time scales from femtoseconds to nanoseconds. It follows that the resonant nonlinear behavior of PT involves many photophysical processes. Singh $et\ al.$, and, more recently, Pang and Prasad concluded, however, that the major part of the NLO response observed with 350fs pulses at 602 nm or with 0fs pulses at 620 nm derives from excitonic phase space filling, and a smaller part is due to polarons.

Logsdon $et\ al.$ have reported that the electrochemically polymerized P3DDT forms a stable monolayer and can be deposited as a LB film. The $\chi^{(3)}$ value of this LB film, measured at 0.602 μm by DFWM, was found to be $\sim 1 \times 10^{-9}$ esu with subpicosecond response. The large $\chi^{(3)}$ value, which is a result of the resonance enhancement, is not surprising when one considers that the optical band gap taken from the onset of the absorption band is ~ 1.9eV.

An analog of PT, poly(isothianaphthene) (PITN) (66a) obtained by benzannelation exhibits the narrowest band gap ($\sim$1eV) among the known conjugated polymers. The NLO behavior of this compound has not been investigated at this point due to its low optical quality. Other modified PTs,polythieno(3,2-b)thiophene (PTT)(66b), polydithien-o(3,2-b, 2′,3′-b)-thiophene (PDTT) (66c) and poly[1,4-di-)2-thienyl)benzene] (PDTB) (66d), have been synthesized by electrochemical polymerization.

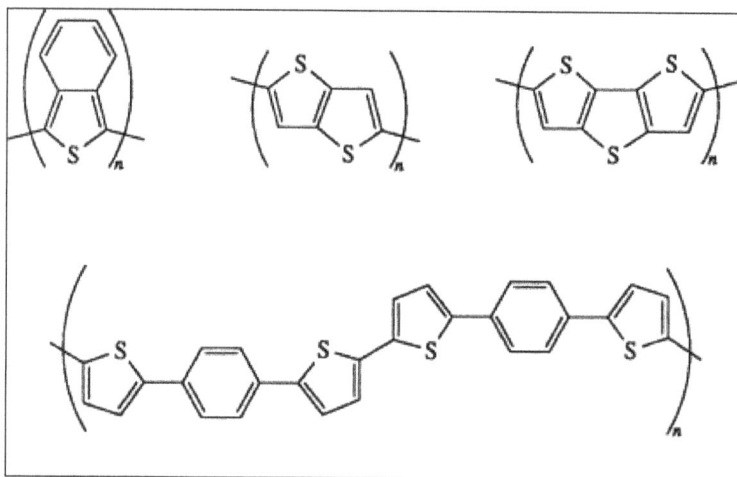

The natural films for PTT and PDTT show good optical quality and their band gaps are both smaller than that of PT. The $\chi^{(3)}$ value measured for PTT and PDTT over the wavelength range 0.53–0.63 µm was determined to be of the order of 10^{-9} esu using DFWM. The large value is expected due to the strong electronic resonance enhancement process.

Luminescent conjugated polythiophenes (LCPs) with molecular scaffolds consisting of polymers of thiophene moieties have evolved as an interesting and uniquely useful class of fluorescent probes. These molecules provide a distinctive optical readout through the impact of detailed biomolecular interactions on the conformation and the geometry of the LCP backbone. The structurally induced optical changes of the LCP backbone also allow the tantalizing prospect of detection of altered conformations of biomolecules. A variety of LCP-based amyloid-specific ligands have been reported. These ligands provide detailed morphology of the protein deposits and conformational phenotyping of distinct and polymorphic protein aggregates. Beyond reporting the total amount, the LCPs allow spectral fingerprinting of polymorphic populations of these assemblies. These ligands have proved effective for studying protein aggregates *ex vivo* and *in vivo* in transgenic mouse models with AD Aβ pathology, in transgenic mice infected with distinct prion strains, and in human clinical tissue samples.

Application of LCPs to transgenic mouse models with AD Aβ pathology revealed a striking heterogeneity in the characteristic plaques composed of the Aβ peptide . LCP staining of brain tissue slices revealed distinct sub-populations of plaques, seen as plaques fluorescing different colors, and intraplaque heterogeneity of color. The spectral features of LCPs are useful for comparison of polymorphic protein aggregates in well-defined experimental systems.

Polymorphic protein aggregates are found in many protein aggregation disorders, and this is especially notable in the infectious prion diseases. Prions can occur as different strains that are most likely encoded in the tertiary or quaternary structure of the prion aggregates. LCP staining of protein aggregates in brain sections from mice infected with distinct prion strains exhibited a distinct spectroscopic signature when bound to strain-specific prion deposits . Different prion strains could also be distinguished by their staining selectivity by LCPs with distinct ionic side chains. The spectral differences due to different thiophene backbone conformations for the anionic LCP, polythiophene acetic acid (PTAA), were visualized by correlation plots of the ratios of the intensities of fluorescence emission at specific wavelengths. Prion aggregates associated with distinct prion strains, chronic wasting disease (CWD), and sheep scrapie, were easily distinguished. LCP fluorescence analysis has also been used to identify novel prion strains and to investigate the molecular basis for interspecies prion disease transmission in mice. PTAA allows ready characterization and identification of mixed prion strains within a single host. The presence of multiple prion strains is difficult to demonstrate by conventional biochemical methods.

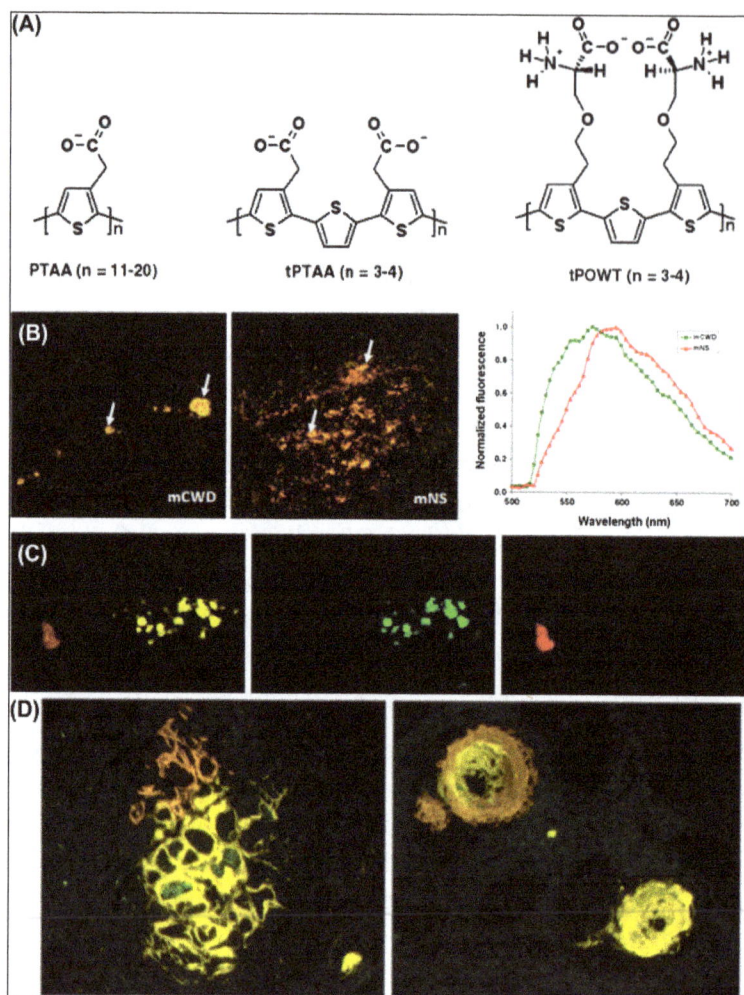

(A)

PTAA (n = 11-20)   tPTAA (n = 3-4)   tPOWT (n = 3-4)

(B)

mCWD   mNS

(C)

(D)

In the above figure, (A) Chemical structures of luminescent conjugated polythiophenes that have been utilized as amyloid ligands. (B) Fluorescence images and spectra (right) of prion deposits, chronic wasting disease (mCWD) (left) and sheep scrapie (mNS) (middle) stained by PTAA. Some representative PTAA-stained prion deposits are indicated by white arrows. (C) Fluorescence images of PTAA stained mCWD and mNS deposits in a single host. mCWD deposits have a green spectrum (middle), whereas mNS deposits have a red spectrum (right). (D) Fluorescence image of PTAA bound to amyloid deposits in tissue samples, heart (left) and liver (right) from a patient diagnosed with AA-amyloidosis. Two different colors are observed from distinct deposits, suggesting the presence of polymorphic AA-amyloid.

Studies of prion deposits in tissue samples suggest that the LCP emission profile is an indirect readout of structural difference between prion deposits associated with distinct prion strains. To mechanistically relate geometrical alterations of the LCPs to structural variations of the protein deposits, rather than their composition, studies were carried out with recombinant prion protein, PrP. Different forms of *in vitro* amyloid fibrils generated from recombinant mouse, Syrian hamster or human prion protein (mPrP, sHaPrP, HuPrP) displayed distinct LCP spectra. The emission profile of PTAA distinguished individual aggregates within different PrP fibril preparations that were chemically identical (same PrP sequence). Thus, spectral differences in tissue observed for PTAA were most likely due to structural differences between the fibrils. LCP fingerprinting will be useful for the analysis of conformational polymorphism in other protein aggregation diseases, as similar phenomena may be much more frequent in other neurodegenerative protein aggregation disorders and amyloidoses . PTAA emission spectral profiles can be used to sub-type systemic amyloidoses in tissue samples similar to prion strains; PTAA staining revealed the existence of multiple types of AA amyloid in a single host.

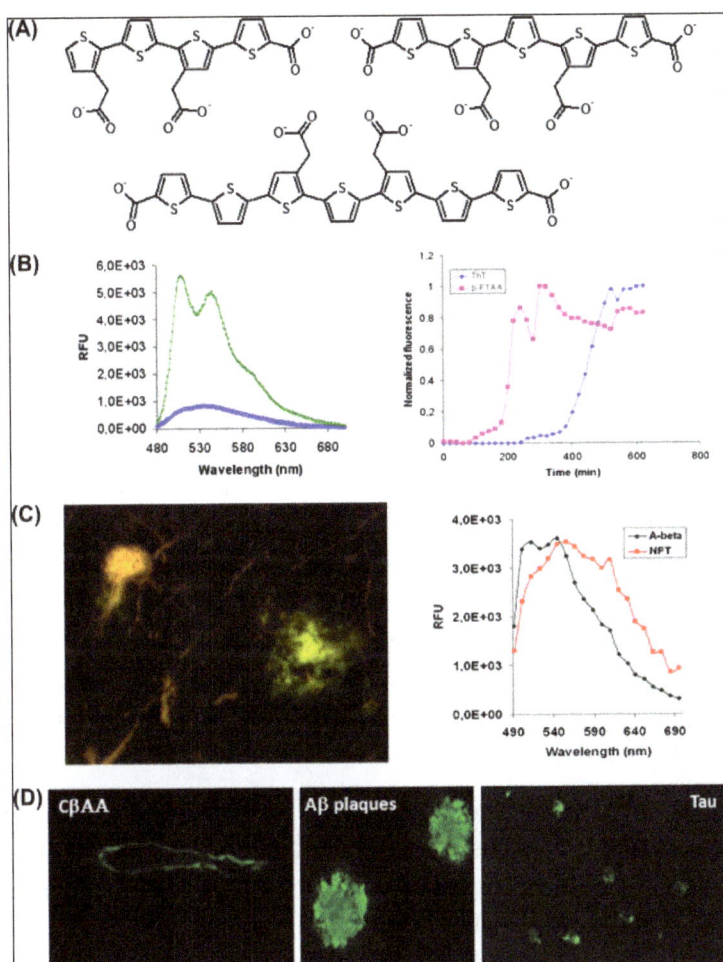

Although LCPs are uniquely suited to spectroscopic characterization of polymorphic amyloid deposits, they do not detect pre-fibrillar aggregates preceding the formation of mature amyloid fibrils. They are also relatively high molecular weight, polydisperse, and do not penetrate the blood–brain barrier for *in vivo* imaging of protein aggregates in the brain. In 2009, Åslund and co-workers introduced a novel class of smaller chemically defined LCPs, denoted luminescent conjugated oligothiophenes (LCOs), based on a pentameric thiophene backbone. These molecules are amyloid-specific ligands under physiologic conditions and show striking enhanced fluorescence and distinct emission wavelength profiles when bound to protein aggregates associated with AD and prion diseases. The LCOs display a distinct emission profile with well-resolved sub-structure upon binding to recombinant Aβ(1-42) amyloid fibrils, implying that the backbone of the LCOs becomes more rigid upon binding to the fibrils. Both thioflavin T (ThT) and one of the LCOs, p-FTAA, reported conventional nucleated fibrillation kinetic behavior of recombinant Aβ(1-42) peptide, including a lag phase, a rapid exponential growth phase, and a final plateau phase. However, with the recombinant Aβ(1-40) peptide, pFTAA fluorescence revealed an earlier growth phase than ThT, indicating that p-FTAA detected pre-fibrillar Aβ(1-40) species preceding the formation of amyloid fibrils. p-FTAA also detects non-thioflavinophilic pre-fibrillar aggregates *in vitro* for a variety of other amyloidogenic proteins, including PrP, insulin, lysozyme and different Aβ peptides. Using a small library of thiophenes with distinct chain length, Nilsson and co-workers were able to show that a thiophene backbone consisting of at least five thiophene units was necessary to detect the pre-fibrillar aggregated species.

In the figure (A) Chemical structures of tetrameric (q-FTAA), pentameric (p-FTAA) and heptameric (h-FTAA) luminescent conjugated oligothiophenes that have been utilized as amyloid ligands. (B) Fluorescence spectrum (left) of p-FTAA mixed with recombinant Aβ(1-40) amyloid fibrils (green) or freshly dissolved Aβ(1-40) (blue). Comparison between ThT and p-FTAA for monitoring the kinetics of recombinant Aβ(1-40) amyloid fibril formation. Notably, p-FTAA reacts earlier than ThT. (C) Fluorescence image (left) and spectra (right) of p-FTAA bound to the two pathologic hallmarks, Aβ deposits (green) and neurofibrillary tangles (yellow-red), in brain tissue samples from an AD patient. (D) Fluorescence images after a single intravenous injection of hFTAA in transgenic mice. Characteristic amyloid lesions, cerebrovascular β-amyloid angiopathy (CβAA), Aβ plaques, and intracellular tau aggregates are labeled throughout the brain and visualized by green fluorescence from h-FTAA.

The high selectivity for protein aggregates and the distinctive conformation-induced optical properties of the novel chemically defined LCOs were further demonstrated when applied to cryo-sectioned brain tissue from AD patients The major pathologic hallmarks of AD, Aβ deposits, neurofibrillary tangles (NFTs) and dystrophic neurites, were clearly detected by all of the LCOs. Moreover, the LCOs showed complete co-localization with 6E10 and AT8, antibodies conventionally used to stain Aβ and phosphorylated tau in NFTs, respectively. Pentameric, hexameric, and heptameric LCOs with the terminal thiophenes substituted with carboxyl groups at the α-position (gave different emission spectra when bound to the Aβ and tau pathologies in the human AD brain. This novel class of oligothiophene-based amyloid-specific dyes is a promising histologic tool for spectral assignment of distinct protein aggregates observed in AD.

Surprisingly, anionic LCOs with four negative charges can be used for *in vivo* optical imaging of protein aggregates. The labeling of plaques in the transgenic APP/PS1 mouse brain can be observed in real time by multiphoton microscopy through a cranial window overlying the parietal cortex. The staining was persistent as individual LCO stained amyloid deposits were detectable even 1 week post-injection of the dye. p-FTAA also specifically labels prion deposits associated with distinct prion strains *in vivo*, and the strains could be distinguished by their p-FTAA spectral signature . Lately, it was also shown that the heptameric LCO, h-FTAA, could be utilized for spectral assignment of CβAA, Aβ plaques and intracellular tau aggregates in transgenic mice after a single intravenous injection of the LCO .

Although the conformation-dependent spectral information would be lost, radiolabeled versions of the oligomeric thiophenes for clinical PET imaging could be developed. Optimization of the thiophene core structure will be required to provide the requisite binding selectivities for the different protein pathologies, the different species of protein aggregates (oligomers, fibrils), and their polymorphic forms.

References

- Difference-between-organic-and-inorganic-polymers: differencebetween.com, Retrieved 25 June, 2019

- McNeill, R.; Siudak, R.; Wardlaw, J. H.; Weiss, D. E. (1963). "Electronic Conduction in PolymersI. The Chemical Structure of Polypyrrole". Aust. J. Chem. 16 (6): 1056–75. doi:10.1071/CH9631056

- Luo, Liang; Wilhelm, Christopher; Young, Christopher N.; Grey, Clare P.; Halada, Gary P.; Xiao, Kai; Ivanov, Ilia N.; Howe, Jane Y.; Geohegan, David B.; Goroff, Nancy S. (2011), "Characterization and Carbonization of Highly Oriented Poly(diiododiacetylene) Nanofibers", Macromolecules, 44 (8): 2626–2631, Bibcode:2011MaMol..44.2626L, doi:10.1021/ma102324r

- Polythiophene, biochemistry-genetics-and-molecular-biology: sciencedirect.com, Retrieved 16 May, 2019

- David Parker, Jan Bussink, Hendrik T. van de Grampel, Gary W. Wheatley, Ernst-Ulrich Dorf, Edgar Ostlinning, Klaus Reinking, "Polymers, High-Temperature" in Ullmann's Encyclopedia of Industrial Chemistry 2002, Wiley-VCH: Weinheim. doi:10.1002/14356007.a21_449

3
Inorganic Polymers

The polymers which do not contain carbon atoms are known as inorganic polymers. Some of the common examples of inorganic polymers are geopolymer, polystannane, polythiazyl and polysilazane. All these different types of inorganic polymers have been carefully analyzed in this chapter.

Inorganic polymer is any of a class of large molecules that lack carbon and are polymers—that is, made up of many small repeating units called monomers. Nature abounds with carbon-based (that is, organic) polymers, such as wool, silk, proteins, starch, and cellulose. In addition, rubber and plastic are made of a wide variety of man-made organic polymers. But many inorganic compounds, such as oxyacids and oxy-anions, also form polymers. This is especially true of weak acids, such as boric acid, H_3BO_3, and silicic acid, H_4SiO_4. In the anions of weak acids, a high density of negative charge resides on the oxygen atoms. This charge density can be reduced by the process of polymerization.

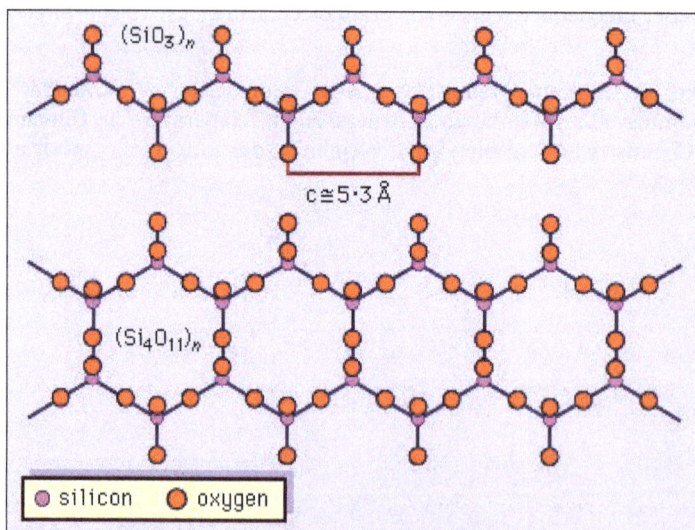

The single-chain silicon-oxygen tetrahedral structure $(SiO_3)n$ of pyroxene minerals and the double-chain structure $(Si_4O_{11})n$ of amphibole minerals are examples of inorganic polymers of silicon.

Borates

These compounds are salts of the oxyacids of boron (B), such as boric acid, H_3BO_3, metaboric acid, HBO_2, and tetraboric acid, $H_2B_4O_7$. Borates result either from the reaction of a base with a boron oxyacid or from the melting of boric acid or boron oxide, B_2O_3, with a molten metal oxide or hydroxide. Borate anion structures range from the simple trigonal planar BO_3^{3-} ion to rather complex structures containing chains and rings of three- and four-coordinated boron atoms. For example, calcium metaborate, CaB_2O_4, consists of infinite chains of $B_2O_4^{2-}$ units, whereas potassium borate, $K[B_5O_6(OH)_4] \cdot 2H_2O$ (commonly written as $KB_5O_8 \cdot 4H_2O$), consists of two B_3O_3 rings linked through a common four-coordinated boron atom. The tetraborates, $B_4O_5(OH)_4^{2-}$, contain both three- and four-coordinated boron surrounded trigonally and tetrahedrally, respectively, by oxygen (O) atoms. Commercially, the most important borate is borax, or sodium tetraborate decahydrate, $Na_2B_4O_7 \cdot 10H_2O$. Borax is found naturally in dry lake beds, such as Searles Lake in California. It can be used to soften water and to make washing compounds. Its usefulness arises from the insolubility of calcium and magnesium borates and the alkaline or basic nature of aqueous solutions of borax. Borax is also used in the manufacture of borosilicate glass and enamels and as a fire retardant.

Silicates

Silicates are salts containing anions of silicon (Si) and oxygen. There are many types of silicates, because the silicon-to-oxygen ratio can vary widely. In all silicates, however, silicon atoms are found at the centres of tetrahedrons with oxygen atoms at the corners. The silicon is always tetravalent (i.e., has an oxidation state of +4). The variation in the silicon-to-oxygen ratio occurs because the silicon-oxygen tetrahedrons may exist as discrete, independent units or may share oxygen atoms at corners, edges, or—in rarer instances—faces in several ways. Thus, the silicon-to-oxygen ratio varies according to the extent to which the oxygen atoms are shared by silicon atoms as the tetrahedrons are linked together. The linkage of these tetrahedrons provides a rather convenient way of classifying silicates. Seven different classifications are commonly recognized.

1. In some silicates, individual SiO_4^{4-} tetrahedrons exist as independent units. Silicates of magnesium (Mg_2SiO_4) and zirconium ($ZrSiO_4$) are examples.

2. Two SiO_4 tetrahedrons share one corner oxygen atom to form discrete $Si_2O_7^{6-}$ ions. Two compounds with this type of linkage are $Ca_2ZnSi_2O_7$ and $Zn_4(OH)_2Si_2O_7 \cdot H_2O$.

3. SiO_4 tetrahedrons may share corners and form closed rings. In $BaTiSi_3O_9$, three SiO_4 tetrahedrons share corners, whereas in $Be_3Al_2Si_6O_{18}$ (beryl, the deep green variety of which is known as emerald), six tetrahedrons share corners to form a closed ring.

4. SiO_4 tetrahedrons in which each tetrahedron shares two oxygen atoms from two other tetrahedrons exist as chains in some silicates. An example of this type of silicate is $CaMg(SiO_3)_2$. From the formula it appears that SiO_3^{2-} ions exist, but these ions do not occur as independent entities. Parallel chains extend the full length of the crystal and are held together by the positively charged metal ions lying between them.

5. When SiO_4 tetrahedrons in single chains share oxygen atoms, double silicon-oxygen chains form. Metal cations link the parallel chains together. Many of these silicates are fibrous in nature, because the ionic bonds between the metal cations and the silicate anions are not as strong as the silicon-oxygen bonds within the chains. A class of fibrous silicate minerals that belong to this group is collectively called asbestos. The best known and most abundant kind of asbestos is chrysotile, which has the formula $Mg_3(Si_2O_5)(OH)_4$. This compound exists as fibres more than 20 mm (0.8 inch) in length. It was used in the past in many fireproofing and insulation applications, but its use for these purposes has been discontinued because prolonged exposure to airborne asbestos fibres may cause lung cancer.

6. When oxygen atoms are shared between double chains, silicon-oxygen sheets are formed. Metal ions form ionic bonds between the sheets. These ionic bonds are weaker than the silicon-oxygen bonds within the sheets, so silicates with this structure cleave into thin layers. An example of this class of silicates includes talc, $Mg_3Si_4O_{10}(OH)_2$.

7. A most interesting class of silicates consists of the zeolites. These compounds are three-dimensional silicon-oxygen networks with some of the tetravalent silicon ions replaced by trivalent aluminum (Al^{3+}) ions. The negative charge that results—because each Al^{3+} ion has one fewer positive charge than the Si^{4+} ion it replaces—is neutralized by a distribution of positive ions throughout the network. An example of a zeolite is $Na_2(Al_2Si_3O_{10}) \cdot 2H_2O$. Zeolites are characterized by the presence of tunnels and systems of interconnected cavities in their structures. Zeolites are used as molecular sieves to remove water and other small molecules from mixtures, and they can also be employed to separate molecules for which the molecular masses are the same or similar but the molecular structures are different. In addition, they are used as solid supports for highly dispersed catalysts and to promote specific size-dependent chemical reactions.

Silicones

Silicones are polymeric organosilicon compounds containing Si−O−Si linkages and Si−C bonds. They are generally very stable, because of the presence of strong silicon-oxygen and silicon-carbon bonds. A general formula for silicones is $(R_2SiO)_x$, where R can be any one of a variety of organic groups. Silicones may be linear, cyclic, or cross-linked polymers, as shown here.

linear silicone

cyclic silicone

cross-linked silicone

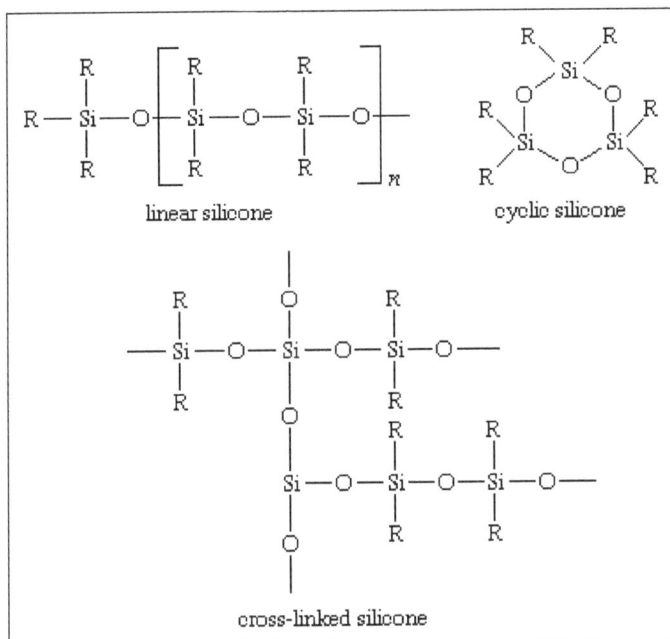

Linear and cyclic silicones are produced by the reaction of water with organochlorosilanes of the general formula R_2SiCl_2, followed by a polymerization reaction that occurs by the elimination of a molecule of water from two hydroxyl groups of adjacent $R_2Si(OH)_2$ molecules.

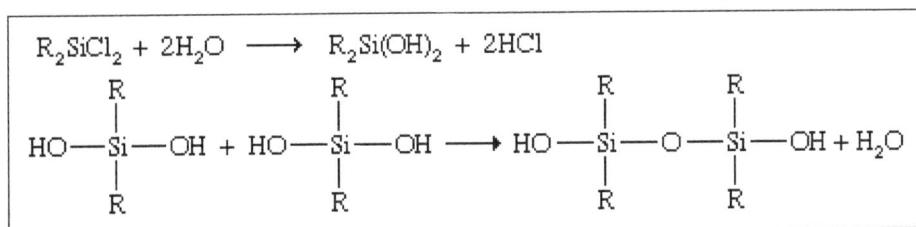

Silicone polymers incorporate some of the properties of both carbon-hydrogen compounds and silicon-oxygen compounds. They are stable to many chemical reagents and to heat. Depending on their degree of polymerization and the complexity of the attached organic groups, silicones can occur in the form of oils, greases, rubberlike substances, or resins. They are used as lubricants, hydraulic fluids, and electrical insulators. They are especially useful as lubricants in applications where there are extreme variations in temperature, because their viscosity changes very little as the temperature changes. Silicones are also water-repellent. Paper, wool, silk, and other fabrics can be coated with a water-repellent film by exposing them for a short time (one to two seconds) to the vapour of trimethylchlorosilane, $(CH_3)_3SiCl$. The $-OH$ groups on the surface of the materials react with the silane, and the surface becomes coated with a thin water-repellent film of $(CH_3)Si-O-$ groups.

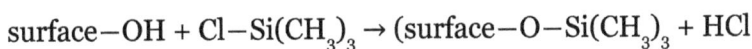

$$surface-OH + Cl-Si(CH_3)_3 \rightarrow (surface-O-Si(CH_3)_3 + HCl$$

Silanes

Silanes are compounds of silicon and hydrogen. Silicon forms a series of hydrides that have the general formula Si_nH_{2n} + 2, including SiH_4, Si_2H_6, Si_3H_8, and Si_4H_{10}. These compounds contain $Si-H$ and $Si-Si$ single bonds. Silicon possesses empty valence-shell d orbitals, which cause the chemistry of silanes to be quite different from that of the corresponding carbon-hydrogen compounds (hydrocarbons). For example, silanes inflame spontaneously in air, whereas the corresponding hydrocarbons do not.

The simplest silane, SiH_4, is called silane. It has a formula and a tetrahedral structure analogous to the hydrocarbon methane, CH_4. It can be prepared in the laboratory by the addition of aqueous acid to an ionic silicide such as magnesium silicide, Mg_2Si.

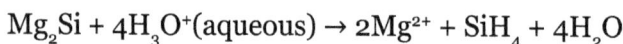

$$Mg_2Si + 4H_3O^+(aqueous) \rightarrow 2Mg^{2+} + SiH_4 + 4H_2O$$

Silane is a colourless gas that is thermally stable at normal temperature but reacts violently with air to produce silicon dioxide and water.

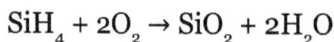

$$SiH_4 + 2O_2 \rightarrow SiO_2 + 2H_2O$$

Silanes react readily with hydrogen halides to produce halogenated silanes. For example, silane reacts in a stepwise manner with hydrogen bromide, HBr, to yield all possible combinations of brominated silane from SiH_3Br to $SiBr_4$. Silanes are also very reactive toward hydroxides, producing silicate ions and hydrogen gas.

$$SiH_4 + 2OH^- + H_2O \rightarrow SiO_3^{2-} + 4H_2$$

In contrast, hydrocarbons are inert to hydroxides. The most important reaction of silanes from a commercial standpoint is their reaction with alkenes—i.e., hydrocarbons that contain a carbon-carbon double bond. For example, the reaction of dichlorosilane, H_2SiCl_2, with propene, $H_3CCH=CH_2$, yields a diorganodihalosilane that can be used in the preparation of silicones.

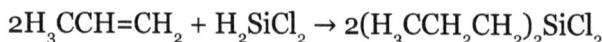

$$2H_3CCH=CH_2 + H_2SiCl_2 \rightarrow 2(H_3CCH_2CH_2)_2SiCl_2$$

Smart Inorganic Polymers

Ibuprofen is encapsulated within the hydrogen bonding network of a smart polysiloxane above LCST, and then released when temperature falls below LCST and the polymer becomes water-soluble.

Smart inorganic polymers (SIPs) are hybrid or fully inorganic polymers with tunable (smart) properties such as stimuli responsive physical properties (shape, conductivity, rheology, bioactivity, self-repair, sensing etc.). While organic polymers are often petrol-based, the backbones of SIPs are made from elements other than carbon which can lessen the burden on scarce non-renewable resources and provide more sustainable alternatives. Common backbones utilized in SIPs include polysiloxanes, polyphosphates, and polyphosphazenes, to name a few.

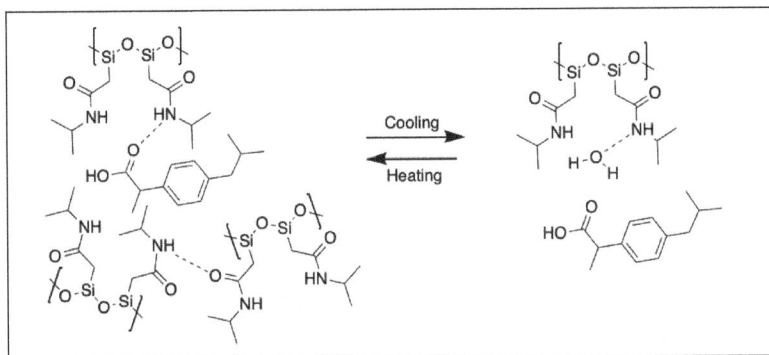

SIPs have the potential for broad applicability in diverse fields spanning from drug delivery and tissue regeneration to coatings and electronics. As compared to organic polymers, inorganic polymers in general possess improved performance and environmental compatibility (no need for plasticizers, intrinsically flame-retardant properties). The unique properties of different SIPs can additionally make them useful in a diverse range of technologically novel applications, such as solid polymer electrolytes for consumer electronics, molecular electronics with non-metal elements to replace metal-based conductors, electrochromic materials, self-healing coatings, biosensors, and self-assembling materials.

Role of COST Action CM1302

COST action 1302 is a European Community "Cooperation in Science and Technology" research network initiative that supported 62 scientific projects in the area of smart inorganic polymers resulting in 70 publications between 2014 and 2018, with the mission of establishing a framework with which to rationally design new smart inorganic polymers. This represents a large share of the total body of work on SIPs. The results of this work are reviewed in the 2019 book, Smart Inorganic Polymers: Synthesis, Properties, and Emerging Applications in Materials and Life Sciences.

Smart Polysiloxanes

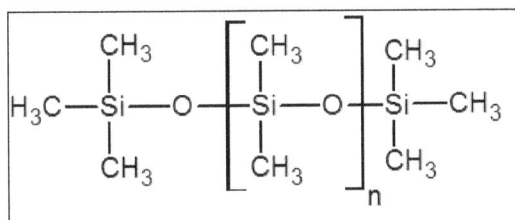

A generic polysiloxane.

Polysiloxane, commonly known as silicone, is the most commonly commercially available inorganic polymer. The large body of existing work on polysiloxane has made it a readily available platform for functionalization to create smart polymers, with a variety of approaches reported which generally center around the addition of metal oxides to a commercially available polysiloxane or the inclusion of functional side-chains on the polysiloxane backbone. The applications of smart polysiloxanes vary greatly, ranging from drug delivery, to smart coatings, to electrochromics.

Drug Delivery

Synthesis of smart stimuli responsive polysiloxanes through the addition of a polysiloxane amine to an α,β-unsaturated carbonyl via aza-Michael addition to create a polysiloxane with N-isopropyl amide side-chains has been reported. This polysiloxane was shown to be able to load ibuprofen (a hydrophobic NSAID) and then release it in response to changes in temperature, showing it to be a promising candidate for smart drug delivery of hydrophobic drugs. This action was attributed to the polymer's ability to retain the ibuprofen above the lower critical solution temperature (LCST), and conversely, to dissolve below the LCST, thus releasing the loaded ibuprofen at a given, known temperature.

Coatings

Commercial polysiloxane coatings are readily commercially available and capable of protecting surfaces from damaging pollutants, but the addition of TiO_2 gives them the smart ability to degrade pollutants stuck to their surface in the presence of sunlight. This particular phenomena is promising in the field of monument preservation. Similar hybrid textile coatings made of amino-functionalized polysiloxane with TiO_2 and silver nanoparticles have been reported to have smart stain-repellent yet hydrophilic properties, making them unique in comparison to typical hydrophobic stain-repellant coatings. Smart properties have also been reported for polysiloxane coatings without metal oxides, namely, a polysiloxane/polyethylenimine coating designed to protect magnesium from corrosion that was found to be capable of self-healing small scratches.

Poly-(ε-caprolactone)/Siloxane

Poly-(ε-caprolactone)/siloxane is an inorganic-organic hybrid material which, when used as a solid electrolyte matrix with a lithium perchlorate electrolyte, paired to a W_2O_3 film, responds to a change in electrical potential by changing transparency. This makes it a potentially useful electrochromic smart glass.

Smart Phosphorus Polymers

There exist a sizable number of phosphorus polymers with backbones ranging from primarily phosphorus to primarily organic with phosphorus subunits. Some of these have been shown to possess smart properties, and are largely of-interest due to the biocompatibility of phosphorus for biological applications like drug delivery, tissue engineering, and tissue repair.

Polyphosphates

Polyphosphate (PolyP) is an inorganic polymer made from phosphate subunits. It typically exists in its deprotonated form, and can form salts with physiological metal cations like Ca^{2+}, Sr^{2+}, and Mg^{2+}. When salted to these metals, it can selectively induce bone regeneration (Ca-PolyP), bone hardening (Sr-PolyP), or cartilage regeneration (Mg-PolyP) depending on the metal to which it is salted. This smart ability to attenuate the kind of tissue regenerated in response to different metal cations makes it a promising polymer for biomedical applications.

Polyphosphazenes

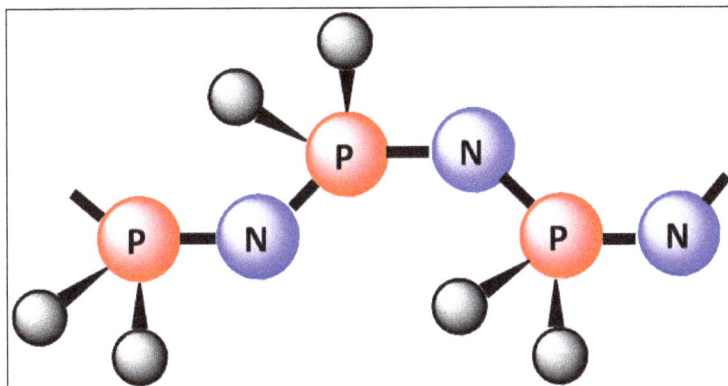

A generic polyphosphazene.

Polyphosphazene is an inorganic polymer with a backbone consisting of phosphorus and nitrogen, which can also form inorganic-organic hybrid polymers with the addition of organic substituents. Some polyphosphazenes have been designed through the addition of amino acid ester side chains such that their LCST is near body temperature and thus they can form a gel *in situ* upon injection into a person, making them potentially useful for drug delivery. They biodegrade into a near-neutral pH mixture of phosphates and ammonia that has been shown to be non-toxic, and the rate of their biodegradation can be tuned with the addition of different substituents from full decomposition within days with glyceryl derivatives, to biostable with fluoroalkoxy substituents.

Poly-ProDOT-Me$_2$

Poly-ProDOT-Me$_2$ is a phosphorus-based inorganic-organic hybrid polymer, which, when paired to a V_2O_5 film, provides a material that changes color upon application of an electrical current. This 'smart glass' is capable of reducing light transmission from

57% to 28% in under 1 second, a much faster transformation than that of commercially available photochromic lenses.

Smart Metalloid and Metal Containing Polymers

While metals are not typically associated with polymeric structures, the inclusion of metal atoms either throughout the backbone of, or as pendant structures on a polymer can provide unique smart properties, especially in relation to their redox and electronic properties. These desirable properties can range from self-repair of oxidation, to sensing, to smart material self-assembly.

Polystannanes

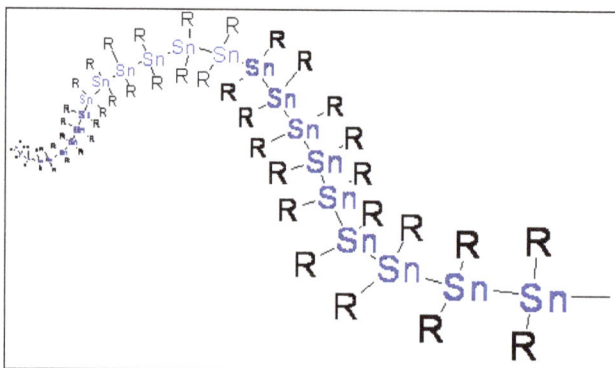

A generic polystannane.

Polystannane, a unique polymer class with a tin backbone, is the only known polymer to possess a completely organometallic backbone. It is especially unique in the way that the conductive tin backbone is surrounded by organic substituents, making it act as an atomic-scale insulated wire. Some polystannanes such as $(SnBu_2)_n$ and $(SnOct_2)_n$ have shown the smart ability to align themselves with external stimuli, which could see them become useful for pico electronics. However, polystannane is very unstable to light, so any such advancement would require a method for stabilizing it against light degradation.

Icosahedral Boron Polymers

Icosahedral boron is a geometrically unusual allotrope of boron, which can be either added as side chains to a polymer or co-polymerized into the backbone. Icosahedral boron side chains on polypyrrole have been shown to allow the polypyrrole to self-repair when overoxidized because the icosahedral boron acts as a doping agent, enabling overoxidation to be reversed.

Polyferrocenylsilane

Polyferrocenylsilanes are a group of common organosilicon metallopolymer with backbones consisting of silicon and ferrocene. Variants of polyferroceylsilanes have been found to exhibit smart self-assembly in response to oxidation and subsequent smart self-disassembly upon reduction, as well as variants which can respond to electrochemical stimulation. One such example is a thin film of a polystyrene-polyferrocenylsilane inorganic-organic hybrid copolymer that was found to be able to adsorb and release ferritin with the application of an electrical potential.

Ferrocene Biosensing

A number of ferrocene-organic inorganic-organic hybrid polymers have been reported to have smart properties that make them useful for application in biosensing. Multiple polymers with ferrocene side-chains cross-linked with glucose oxidase have shown oxidation activity which results in electrical potential in the presence of glucose, making them useful as glucose biosensors. This sort of activity is not limited to glucose, as other enzymes can be crosslinked to allow for sensing of their corresponding molecules, like a poly(vinylferrocene)/carboxylated multiwall carbon nanotube/gelatin composite that was bound to uricase, giving it the ability to act as a biosensor for uric acid.

Geopolymer

Geopolymers are inorganic, typically ceramic, materials that form long-range, covalently bonded, non-crystalline (amorphous) networks. Obsidian (volcanic glass) fragments are a component of some geopolymer blends. Commercially produced geopolymers may be used for fire- and heat-resistant coatings and adhesives, medicinal applications, high-temperature ceramics, new binders for fire-resistant fiber composites, toxic and radioactive waste encapsulation and new cements for concrete. The properties and uses of geopolymers are being explored in many scientific and industrial disciplines: modern inorganic chemistry, physical chemistry, colloid chemistry, mineralogy, geology, and in other types of engineering process technologies. Geopolymers are part of polymer science, chemistry and technology that forms one of the major areas of materials science. Polymers are either organic material, i.e. carbon-based, or inorganic polymer, for example silicon-based. The organic polymers comprise the classes of natural polymers (rubber, cellulose), synthetic organic polymers (textile fibers, plastics, films, elastomers, etc.) and natural biopolymers (biology, medicine, pharmacy). Raw materials used in the synthesis of silicon-based polymers are mainly rock-forming minerals of geological origin, hence the name: *geopolymer*. Joseph Davidovits coined the term in 1978 and created the non profit French scientific institution (Association Loi 1901) *Institut Géopolymère* (Geopolymer Institute).

According to T.F. Yen geopolymers can be classified into two major groups: pure inorganic geopolymers and organic containing geopolymers, synthetic analogues of naturally occurring macromolecules. In the following presentation, a geopolymer is essentially a mineral chemical compound or mixture of compounds consisting of re-peating units, for example silico-oxide (-Si-O-Si-O-), silico-aluminate (-Si-O-Al-O-), ferro-silico-aluminate (-Fe-O-Si-O-Al-O-) or alumino-phosphate (-Al-O-P-O-), created through a process of geopolymerization. This mineral synthesis (geosynthesis) was first presented at an IUPAC symposium in 1976.

The microstructure of geopolymers is essentially temperature dependent:

- It is X-ray amorphous at room temperature,

- But evolves into a crystalline matrix at temperatures above 500 °C.

One can distinguish between two synthesis routes:

- In alkaline medium (Na^+, K^+, Li^+, Ca^{2+}, Cs^+ and the like),

- In acidic medium with phosphoric acid, organic carboxylic acids from plant ex-tracts (acetic, citric, oxalic, and humic acids).

The alkaline route is the most important in terms of research and development and commercial applications and will be described.

In the 1950s, Viktor Glukovsky, of Kiev, USSR, developed concrete materials originally known under the names "soil silicate concretes" and "soil cements", but since the in-troduction of the geopolymer concept by Joseph Davidovits, the terminology and defi-nitions of 'geopolymer' have become more diverse and often conflicting. The examples below were taken from 2011 scientific publications, written by scientists with different backgrounds.

Various Definitions of the Term Geopolymer

For chemists:

> 'Geopolymers consist of a polymeric Si–O–Al framework, similar to zeo-lites. The main difference to zeolite is geopolymers are amorphous instead of crystalline. The microstructure of geopolymers on a nanometer scale observed by TEM comprises small aluminosilicate clusters with pores dis-persed within a highly porous network. The clusters sizes are between 5 and 10 nanometers.'

For geopolymer material chemists:

> 'The reaction produces SiO_4 and AlO_4, tetrahedral frameworks linked by shared oxygens as poly(sialates) or poly(sialate–siloxo) or poly(sialate–disiloxo) de-pending on the SiO_2/Al_2O_3 ratio in the system. The connection of the tetrahe-dral frameworks is occurred via long-range covalent bonds. Thus, geopolymer

structure is perceived as dense amorphous phase consisting of semi-crystalline 3-D alumino-silicate microstructure.'

For alkali-cement scientists:

'Geopolymers are framework structures produced by condensation of tetrahedral aluminosilicate units, with alkali metal ions balancing the charge associated with tetrahedral Al. Conventionally, geopolymers are synthesized from a two-part mix, consisting of an alkaline solution (often soluble silicate) and solid aluminosilicate materials. Geopolymerization occurs at ambient or slightly elevated temperature, where the leaching of solid aluminosilicate raw materials in alkaline solutions leads to the transfer of leached species from the solid surfaces into a growing gel phase, followed by nucleation and condensation of the gel phase to form a solid binder.'

For geopolymer ceramic chemists:

'Although geopolymer is generally X-ray amorphous if cured at standard pressures and temperatures, it will convert into crystalline ceramic phases like leucite or pollucite upon heating.'

For ceramic scientists:

'Geopolymers are a class of totally inorganic, alumino-silicate based ceramics that are charge balanced by group I oxides. They are rigid gels, which are made under relatively ambient conditions of temperature and pressure into near-net dimension bodies, and which can subsequently be converted to crystalline or glass-ceramic materials.'

Geopolymer Synthesis

Ionic Coordination or Covalent Bonding?

In 1937, W. L. Bragg published a method for classifying all kinds of silicates and their crystal structures based on the concept of the ionic theory by Linus Pauling. The fundamental unit is a tetrahedral complex consisting of a small cation such as Si^{4+}, or Al^{3+} in tetrahedral coordination with four oxygens (Pauling's first rule). Many textbooks explain the geometry of the SiO_4^{4-} tetrahedron and other mineral structures as determined by the relative sizes of the different ions.

This ionic coordination representation is no longer adapted to the requirements of geopolymer chemistry that is governed by covalent bonding mechanisms. The differences between the ionic concept (coordination) and the covalent bonding are profound. The double tetrahedron structure (coordination) is sharing one oxygen anion O^{2-}, whereas in the Si-O-Si- molecular structure, the covalent bond is achieved through Si and O co-sharing only one electron. This results in stronger bond within the latter structure.

The American mineralogist and geochemist G. V. Gibbs and his team studied the polymeric bond Si-O-Si-O and stated in 1982-2000: "The successful modeling of the properties and structures of silica lends credence to the statement that a silica polymorph like quartz can be viewed as a giant molecule bound together by essentially the same forces that bind the atoms of the Si-O-Si skeleton into a small siloxane molecule." The term giant molecule used by G.V. Gibbs is equivalent to the definition of geopolymer and the wording small siloxane molecule describes the actual oligomers of organo-silicon compounds well known as silicone polymer. These siloxane oligomers have the same structure as the silico-aluminate oligomers.

Geopolymerization Starts with Oligomers

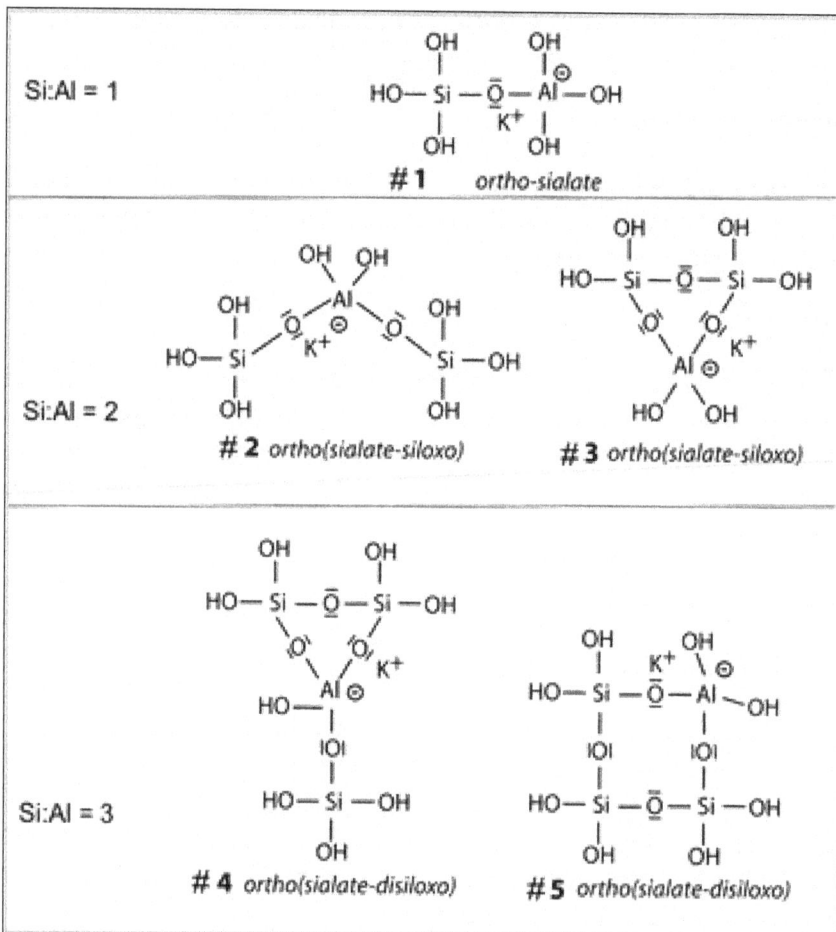

Five isolated oligomers of the K-poly(sialate)/poly(sialate-silxo) species.

Geopolymerization is the process of combining many small molecules known as oligomers into a covalently bonded network. The geo-chemical syntheses are carried out through oligomers (dimer, trimer, tetramer, pentamer) which provide the actual unit structures of the three-dimensional macromolecular edifice. In 2000, T.W. Swaddle and his team proved the existence of soluble isolated alumino-silicate molecules in solution in relatively high concentrations and high pH. One major improvement in their research was that their study was carried out at very low temperatures, as low as −9 °C. Indeed, it was discovered that the polymerization at room temperature of oligo-sialates was taking place on a time scale of around 100 milliseconds, i.e. 100 to 1000 times faster than the polymerization of ortho-silicate, oligo-siloxo units. At room temperature or higher, the reaction is so fast that it cannot be detected with conventional analytical equipment.

The image shows 5 soluble oligomers of the K-poly(sialate)/poly(sialate-siloxo) species, which are the actual starting units of potassium-based alumino-silicate geopolymerization.

Example of (-Si-O-Al-O-) Geopolymerization with Metakaolin MK-750 in Alkaline Medium

It involves four main phases comprising seven chemical reaction steps:

- Alkaline depolymerization of the poly(siloxo) layer of kaolinite.

- Formation of monomeric and oligomeric species, including the "ortho-sialate" $(OH)_3$-Si-O-Al-$(OH)_3$ molecule.

- In the presence of waterglass (soluble K-polysiloxonate), one gets the creation of ortho-sialate-disiloxo cyclic structure, whereby the hydroxide is liberated by condensation reactions and can reacts again.

- Geopolymerization (polycondensation) into higher oligomers and polymeric 3D-networks.

The geopolymerization kinetics for Na-poly(sialate-siloxo) and K-poly(sialate-siloxo) are slightly different respectively. This is probably due to the different dimensions of the Na^+ and K^+ cations, K^+ being bigger than Na^+.

Example of Zeolitic (Si-O-Al-O-) Geopolymerization with Fly Ash in Alkaline Medium

It involves 5 main phases:

- Nucleation stage in which the aluminosilicates from the fly ash particle dissolve in the alkaline medium (Na^+), releasing aluminates and silicates, probably as monomers.

- These monomers inter-react to form dimers, which in turn react with other monomers to form trimers, tetramers and so on.

- When the solution reaches saturation, an aluminum-rich gel (denominated Gel 1) precipitates.

- As the reaction progresses, more Si-O groups from the initial solid source dissolve, increasing the silicon concentration in the medium and gradually raising the proportion of silicon in the zeolite precursor gel (Gel 2).

- Polycondensation into zeolite-like 3D-frameworks.

Geopolymer 3D-frameworks

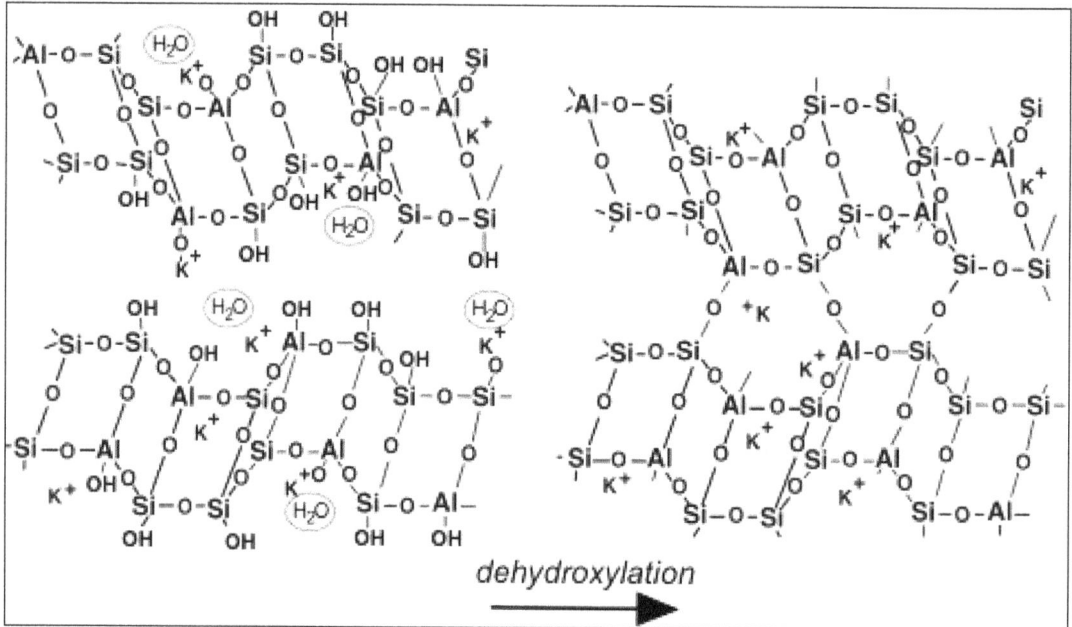

Dehydroxylation of poly(sialate-siloxo) into 3D-framework.

Geopolymerization forms aluminosilicate frameworks that are similar to those of rock-forming minerals. Yet, there are major differences. In 1994, Davidovits presented a theoretical structure for K-poly(sialate-siloxo) (K)-(Si-O-Al-O-Si-O) that was consistent with the NMR spectra. It does not show the presence of water in the structure because he only focused on the relationship between Si, Al, Na, K, atoms. Water is present only at temperatures below 150 °C – 200 °C, essentially in the form of -OH groups, whereas numerous geopolymer industrial and commercial applications work at temperatures above 200 °C, up to 1400 °C, i.e. at temperatures above dehydroxylation. Nevertheless, scientists working on low temperature applications, such as cements and waste management, tried to pinpoint cation hydration and water molecules. This model shows an incompletely reacted geopolymer, which involves free Si-OH groups that will later with time or with temperature polycondense with opposed Al-O-K, into Si-O-Al-O sialate bonds. The water released by this reaction either remains in the pores, is associated with the framework similarly to zeolitic water, or can be released and removed. Several 3D-frameworks are described in the book 'Geopolymer Chemistry and Applications'. After dehydroxylation (and dehydration), generally above 250 °C, geopolymers become more and more crystalline and above 500-1000

°C (depending on the nature of the alkali cation present) crystallise and have X-ray diffraction patterns and framework structures identical to their geological analogues.

Commercial Applications

There exist a wide variety of potential and existing applications. Some of the geopolymer applications are still in development whereas others are already industrialized and commercialized. They are listed in three major categories:

Geopolymer Resins and Binders

- Fire-resistant materials, thermal insulation, foams.

- Low-energy ceramic tiles, refractory items, thermal shock refractories.

- High-tech resin systems, paints, binders and grouts.

- Bio-technologies (materials for medicinal applications).

- Foundry industry (resins), tooling for the manufacture of organic fiber composites.

- Composites for infrastructures repair and strengthening, fire-resistant and heat-resistant high-tech carbon-fiber composites for aircraft interior and automobile.

- Radioactive and toxic waste containment.

Geopolymer Cements and Concretes

- Low-tech building materials (clay bricks).

- Low-CO_2 cements and concretes.

Arts and Archaeology

- Decorative stone artifacts, arts and decoration.

- Cultural heritage, archaeology and history of sciences.

Geopolymer Resins and Binders

The class of geopolymer materials is described by Davidovits to comprise:

- Metakaolin MK-750-based Geopolymer Binder.

  Chemical formula (Na,K)-(Si-O-Al-O-Si-O-), ratio Si:Al=2 (range 1.5 to 2.5).

- Silica-based Geopolymer Binder.

Chemical formula (Na,K)-n(Si-O-)-(Si-O-Al-), ratio Si:Al>20 (range 15 to 40).

- Sol-gel-based Geopolymer Binder (synthetic MK-750).

Chemical formula (Na,K)-(Si-O-Al-O-Si-O-), ratio Si:Al=2.

The first geopolymer resin was described in a French patent application filed by J. Davidovits in 1979. The American patent, US 4,349,386, was granted on Sept. 14, 1982 with the title *Mineral Polymers and methods of making them*. It essentially involved the geopolymerization of alkaline soluble silicate [waterglass or (Na,K)-polysiloxonate] with calcined kaolinitic clay (later coined metakaolin MK-750 to highlight the importance of the temperature of calcination, namely 750 °C in this case). In 1985, Kenneth MacKenzie and his team from New-Zealand, discovered the Al(V) coordination of calcined kaolinite (MK-750), describing a "chemical shift intermediate between tetrahedral and octahedral." This had a great input towards a better understanding of its geopolymeric reactivity. Since 1979, a variety of resins, binders and grouts were developed by the chemical industry, worldwide.

Potential Utilization for Geopolymer Composites Materials

Metakaolin MK-750-based and silica-based geopolymer resins are used to impregnate fibers and fabrics to obtain geopolymer matrix-based fiber composites. These products are fire-resistant; they release no smoke and no toxic fumes. They were tested and recommended by major international institutions such as the American Federal Aviation Administration FAA. FAA selected the carbon-geopolymer composite as the best candidate for the fire-resistant cabin program. Geopolymers are attractive host materials to immobilise nuclear waste due to their high environmental durability and flexibility to compositional changes of waste. They are already used on industrial scale to immobilise difficult radioactive waste streams in Czech Republic and Slovkia.

Fire-resistant Material

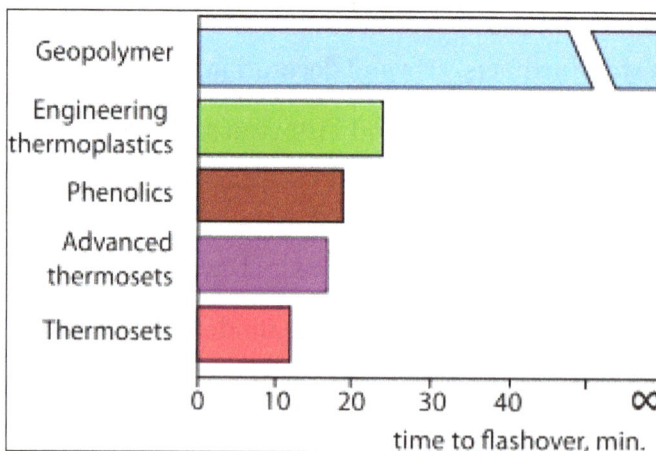

Time to flashover: comparison between organic-matrix and geopolymer-matrix composites.

Flashover is a phenomenon unique to compartment fires where incomplete combustion products accumulate at the ceiling and ignite causing total involvement of the compartment materials and signaling the end to human survivability. Consequently, in a compartment fire the time to flashover is the time available for escape and this is the single most important factor in determining the fire hazard of a material or set of materials in a compartment fire. The Federal Aviation Administration has used the time-to-flashover of materials in aircraft cabin tests as the basis for a heat release and heat release rate acceptance criteria for cabin materials for commercial aircraft. The figure shows how the best organic-matrix made of engineering thermoplastics reaches flashover after the 20 minute ignition period and generates appreciable smoke, while the geopolymer-matrix composite will never ignite, reach flashover, or generate any smoke in a compartment fire.

Carbon-geopolymer composite is applied on racing cars around exhaust parts. This technology could be transferred and applied for the mass production of regular automobile parts (corrosion-resistant exhaust pipes and the like) as well as heat shields. A well-known motorcar manufacturer already developed a geopolymer-composite exhaust pipe system.

Geopolymer Cements

Production of geopolymer cement requires an aluminosilicate precursor material such as metakaolin or fly ash, a user-friendly alkaline reagent (for example, sodium or potassium soluble silicates with a molar ratio MR SiO_2:$M_2O \geq 1.65$, M being Na or K) and water. Room temperature hardening is more readily achieved with the addition of a source of calcium cations, often blast furnace slag.

Portland Cement Chemistry vs Geopolymer Chemistry

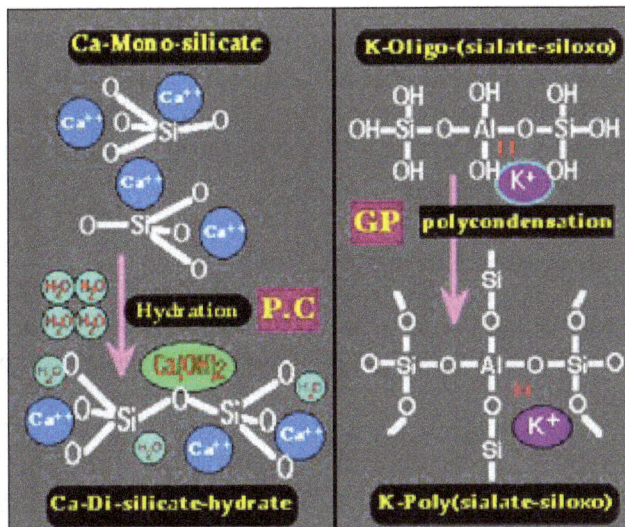

Portland cement chemistry compared to geopolymerization GP.

On the left of the image is shown, hardening of Portland cement (P.C.) through hydration of calcium silicate into calcium silicate hydrate (C-S-H) and portlandite, $Ca(OH)_2$.

On the right of the image is shown, hardening (setting) of geopolymer cement (GP) through poly-condensation of potassium oligo-(sialate-siloxo) into potassium poly(sialate-siloxo) cross linked network.

Geopolymer Cement Categories

The categories comprise:

- Slag-based geopolymer cement.

- Rock-based geopolymer cement.

- Fly ash-based geopolymer cement:

  o Type 1: Alkali-activated fly ash geopolymer.

  o Type 2: Slag/fly ash-based geopolymer cement.

- Ferro-sialate-based geopolymer cement.

Slag-based Geopolymer Cement

Components: Metakaolin (MK-750) + blast furnace slag + alkali silicate (user-friendly).

Geopolymeric make-up: Si:Al = 2 in fact solid solution of Si:Al=1, Ca-poly(di-sialate) (anorthite type) + Si:Al = 3, K-poly(sialate-disiloxo) (orthoclase type) and C-S-H Ca-silicate hydrate.

The first geopolymer cement developed in the 1980s was of the type (K,Na,Ca)-poly(-sialate) (or slag-based geopolymer cement) and resulted from the research developments carried out by Joseph Davidovits and J.L. Sawyer at Lone Star Industries, USA and yielded the invention of Pyrament cement. The American patent application was filed in 1984 and the patent US 4,509,985 was granted on April 9, 1985 with the title 'Early high-strength mineral polymer'.

Rock-based Geopolymer Cement

The replacement of a certain amount of MK-750 with selected volcanic tuffs yields geopolymer cement with better properties and less CO_2 emission than the simple slag-based geopolymer cement.

Manufacture components: Metakaolin MK-750, blast furnace slag, volcanic tuffs (calcined or not calcined), mine tailings and alkali silicate (user-friendly).

Geopolymeric make-up: Si:Al = 3, in fact solid solution of Si:Al = 1 Ca-poly(di-sialate) (anorthite type) + Si:Al = 3-5 (Na,K)-poly(silate-multisiloxo) and C-S-H Ca-silicate hydrate.

Fly Ash-based Geopolymer Cements

Later on, in 1997, building on the works conducted on slag-based geopolymeric cements, on the one hand and on the synthesis of zeolites from fly ashes on the other hand, Silverstrim et al. and van Jaarsveld and van Deventer developed geopolymeric fly ash-based cements.

CO_2 Emissions during Manufacture

According to the Australian concrete expert B. V. Rangan, the growing worldwide demand for concrete is a great opportunity for the development of geopolymer cements of all types, with their much lower tally of carbon dioxide CO_2.

Need for Standards

In June 2012, the institution ASTM International organized a symposium on Geopolymer Binder Systems. The introduction to the symposium states: "When performance specifications for Portland cement were written, non-portland binders were uncommon...New binders such as geopolymers are being increasingly researched, marketed as specialty products, and explored for use in structural concrete. This symposium is intended to provide an opportunity for ASTM to consider whether the existing cement standards provide, on the one hand, an effective framework for further exploration of geopolymer binders and, on the other hand, reliable protection for users of these materials."

The existing Portland cement standards are not adapted to geopolymer cements. They must be created by an *ad hoc* committee. Yet, to do so, requires also the presence of standard geopolymer cements. Presently, every expert is presenting his own recipe based on local raw materials (wastes, by-products or extracted). There is a need for selecting the right geopolymer cement category. The 2012 State of the Geopolymer R&D, suggested to select two categories, namely:

- Type 2 slag/fly ash-based geopolymer cement: Fly ashes are available in the major emerging countries.

- Ferro-sialate-based geopolymer cement: This geological iron rich raw material is present in all countries throughout the globe.

- The appropriate user-friendly geopolymeric reagent.

Geopolymer Applications to Arts and Archaeology

Because geopolymer artifacts look like natural stone, several artists started to cast in silicone rubber molds replications of their sculptures. For example, in the 1980s, the French artist Georges Grimal worked on several geopolymer castable stone formulations.

Egyptian Pyramid Stones

With respect to archaeological applications, in the mid-1980s, Joseph Davidovits presented his first analytical results carried out on genuine pyramid stones. He claimed that the ancient Egyptians knew how to generate a geopolymeric reaction in the making of a re-agglomerated limestone blocks. The Ukrainian scientist G.V. Glukhovsky endorsed Davidovits' research in his keynote paper to the First Intern. Conf. on Alkaline Cements and Concretes, Kiev, Ukraine, 1994. Later on, several materials scientists and physicists took over these archaeological studies and are publishing their results, essentially on pyramid stones.

Roman Cements

From the digging of ancient Roman ruins, one knows that approximately 95% of the concretes and mortars constituting the Roman buildings consist of a very simple lime cement, which hardened slowly through the precipitating action of carbon dioxide CO_2, from the atmosphere and formation of calcium silicate hydrate (C-S-H). This is a very weak to medium good material that was used essentially in the making of foundations and in buildings for the populace.

But for the building of their "ouvrages d'art", especially works related to water storage (cisterns, aqueducts), the Roman architects did not hesitate to use more sophisticated and expensive ingredients. These outstanding Roman cements are based on the calcic activation of ceramic aggregates and alkali rich volcanic tuffs (cretoni, zeolitic pozzolan), respectively with lime. MAS-NMR Spectroscopy investigations were carried out on these high-tech Roman cements dating to the 2nd century AD. They show their geopolymeric make-up.

Polystannane

Polystannanes are organotin compounds with the formula $(R_2Sn)_n$. These polymers have been of intermittent academic interest; they are unusual because heavy elements comprise the backbone. Structurally related but better characterized (and more useful) are the polysilanes $(R_2Si)_n$.

Linear Polystannanes

Dialkytin dihydrides (R_2SnH_2) were reported in 2005 to undergo dehydropolymerization in the presence of Wilkinson's catalyst. This method afforded polystannanes without detectable amounts of "cyclic"-byproducts. The polymers were yellow with number average molar masses of 10 to 70 kg/mol and a polydispersity of 2 – 3. By variation of the catalyst concentration the molar masses of the synthesized polymers could be adjusted. A strong influence of the temperature on the degree of conversion was observed. Determination of the molar mass at different degrees of conversion indicated that polymerization did not proceed according to a statistical condensation mechanism, but, likely, by growth onto the catalyst, e.g. by insertion of SnR_2-like units.

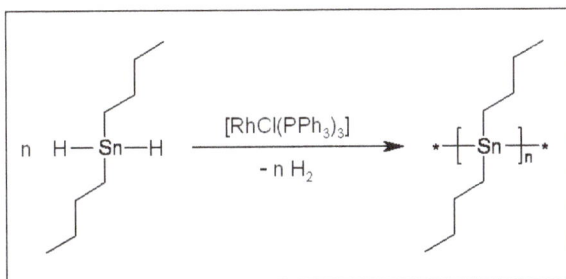

Synthesis of pure linear poly(dibutylstannane).

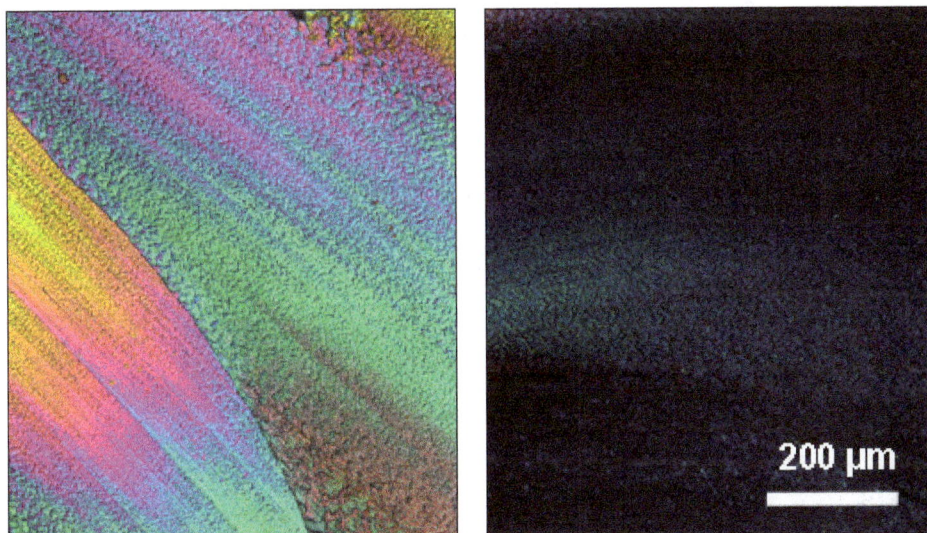

Optical micrographs (crossed polarizers) of an oriented film of poly(3-metylbutylstannane) produced by shearing the material at room temperature, top at 45° and bottom 90° in respect to the polarizers.

The poly(dialkylstannane)s were found to be thermotropic and displayed first-order phase transitions from one liquid-crystalline phase into another or directly to the isotropic state, depending on the length of the side groups. More specifically, poly(dibutylstannane) for example showed an endothermic phase transition at ~0 °C from a rectangular to a pure nematic phase, as determined by X-ray diffraction.

As expected, polystannanes were semi-conductive. Temperature-dependent, time-resolved pulse radiolysis microwave conductivity measurements of poly(dibutylstannane) yielded values of charge-carrier mobilities of 0.1 to 0.03 cm² V⁻¹ s⁻¹, which are similar to those found for pi-bond-conjugated carbon-based polymers.

By partial oxidation of the material with SbF_5 conductivities of 0.3 S cm^{-1} could be monitored.

The liquid-crystalline characteristics of the poly(dialkylstannane)s permitted facile orientation of these macromolecules, for instance, by mechanical shearing or tensile drawing of blends with poly(ethylene). Poly(dialkylstannane)s with short side groups invariably arranged parallel to the external orientation direction, while the polymers with longer side groups had a tendency to order themselves perpendicular to that axis.

Polythiazyl

Polythiazyl (polymeric sulfur nitride), $(SN)_x$, is an electrically conductive, gold- or bronze-colored polymer with metallic luster. It was the first conductive inorganic polymer discovered and was also found to be a superconductor at very low temperatures (below 0.26 K). It is a fibrous solid, described as "lustrous golden on the faces and dark blue-black", depending on the orientation of the sample. It is air stable and insoluble in all solvents. It was first observed in 1910.

Structure and Bonding

The material is a polymer. The S and N atoms on adjacent chains align. Several resonance structures can be written.

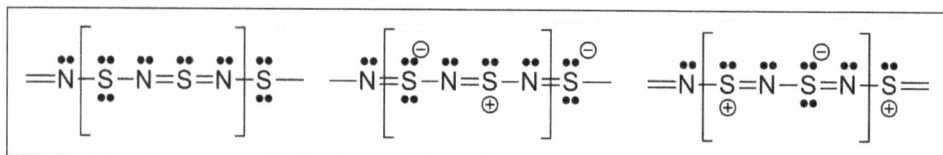

Synthesis

Polythiazyl is synthesized by the polymerization of the dimer disulfur dinitride (S_2N_2), which is in turn synthesized from the cyclic alternating tetramer tetrasulfur tetranitride (S_4N_4). Conversion from cyclic tetramer to dimer is catalysed with hot silver wool.

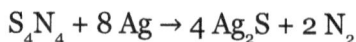

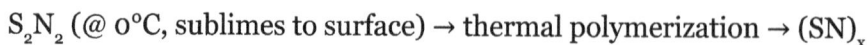

$S_4N_4 + 8\ Ag \rightarrow 4\ Ag_2S + 2\ N_2$

S_4N_4 (w/ Ag_2S catalyst) $\rightarrow$ 2 S_2N_2 (w/ 77K cold finger) $\rightarrow$ S_2N_2

S_2N_2 (@ 0°C, sublimes to surface) $\rightarrow$ thermal polymerization $\rightarrow$ $(SN)_x$

Uses

Due to its electrical conductivity, polythiazyl is used in LEDs, transistors, battery cathodes, and solar cells.

Polysilazane

Polysilazanes are a class of very heat stable inorganic polymers with a polymer backbone made up entirely of silicon-nitrogen bonds with either only hydrogen substituents attached to each silicon and nitrogen atom (perhydro-polysilazanes PHPS, -NH-SiH$_2$-) or with additional organic substituents attached to each silicone (organo polysilazanes, OPSZ, -NH-SiR$_2$-). The pendent groups are typically hydrogen, alkyl, and/or aryl groups. Thus an almost unlimited number of different polymers belong to this class. However, only a small number if silazane monomer/polymers are commercially available.

Polysilazanes are known for their high temperature and oxidative stability; excellent scratch, abrasion and impact resistance; high hardness (up to 5H for OPSZ and 9H for PHPS); and high resistance to weathering and many chemicals. Most or all of these properties are superior to those of polysiloxanes. Polysilazanes have also outstanding non-stick and easy clean properties which are almost as as good as those of fluoropolymers (PTFE).

Polysilazanes are mainly used as precursors to SiO$_2$, Si$_3$N$_4$ and SiC ceramics. Pyrolysis/oxidation of these precursors provide a convenient method for preparing silicon-based ceramic products, such as fibers, coatings and fiber-reinforced ceramic-matrix composites (CMCs) which are often difficult to manufacture with conventional ceramic processing methods. However, the use of polysilazanes has been severely limited by the high price and by lack of availability. The most important applications appear to be protective and heat resistant coatings in the electronic, automotive, and aerospace industry (exhaust systems, engine parts, heat exchangers, ovens and electronic devices). Organo silazanes are also sometimes incorporated into one- and two-component coatings for car/train exteriors and architectural elements. These products protect smooth and non-absorbent surfaces against graffiti and dirt.

References

- Inorganic-polymers-1462212: britannica.com, Retrieved 19 August, 2019

- Caminade AM, Hey-Hawkins E, Manners I (September 2016). "Smart Inorganic Polymers" (PDF). Chemical Society Reviews. 45 (19): 5144–5146. doi:10.1039/C6CS90086K. PMID 27711697

- Silazanes, Polymer-Brands: polymerdatabase.com, Retrieved 26 June, 2019

- Ronald D. Archer (26 February 2001). Inorganic and Organometallic Polymers. John Wiley & Sons. p. 213. ISBN 978-0-471-24187-4. Retrieved 29 June 2012

4
Polymerization

The process of reaction between monomer molecules in a chemical reaction in order to form three-dimensional network or polymer chains is referred to as polymerization. The major types of polymerization are coordination polymerization, dispersion polymerization, step-growth polymerization, nitroxide-mediated radical polymerization, ring-opening polymerization, etc. The topics elaborated in this chapter will help in gaining a better perspective about these types of polymerization.

Polymerization is any process in which relatively small molecules, called monomers, combine chemically to produce a very large chainlike or network molecule, called a polymer. The monomer molecules may be all alike, or they may represent two, three, or more different compounds. Usually at least 100 monomer molecules must be combined to make a product that has certain unique physical properties—such as elasticity, high tensile strength, or the ability to form fibres—that differentiate polymers from substances composed of smaller and simpler molecules; often, many thousands of monomer units are incorporated in a single molecule of a polymer. The formation of stable covalent chemical bonds between the monomers sets polymerization apart from other processes, such as crystallization, in which large numbers of molecules aggregate under the influence of weak intermolecular forces.

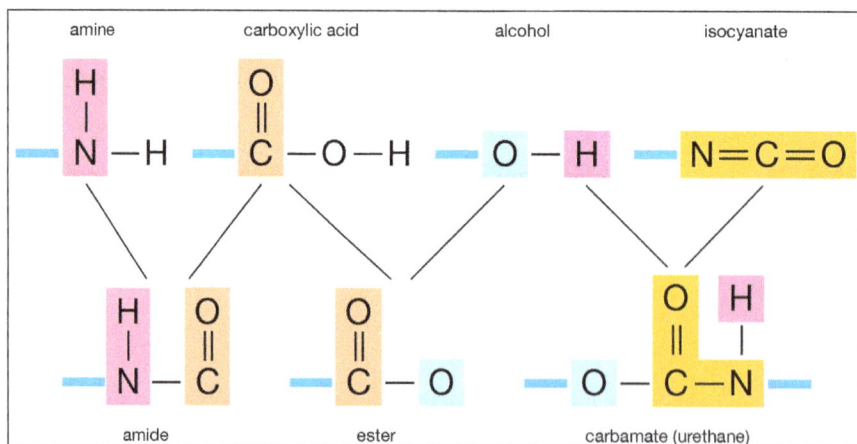

Functional groups in monomers and polymers.

Two classes of polymerization usually are distinguished. In condensation polymeriza-tion, each step of the process is accompanied by the formation of a molecule of some simple compound, often water. In addition polymerization, monomers react to form a polymer without the formation of by-products. Addition polymerizations usually are carried out in the presence of catalysts, which in certain cases exert control over struc-tural details that have important effects on the properties of the polymer.

Linear polymers, which are composed of chainlike molecules, may be viscous liquids or solids with varying degrees of crystallinity; a number of them can be dissolved in certain liquids, and they soften or melt upon heating. Cross-linked polymers, in which the molecular structure is a network, are thermosetting resins (i.e., they form under the influence of heat but, once formed, do not melt or soften upon reheating) that do not dissolve in solvents. Both linear and cross-linked polymers can be made by either addition or condensation polymerization.

Degenerative Chain Transfer

In polymer chemistry, degenerative chain transfer (also called degenerate chain trans-fer) is a process that can occur in a radical polymerization whereby reactivity of active centres are changed, hence significantly influencing the molecular weight distribution of the resulting product.

In chain polymerization processes it is observed that during the chemical reactions an active centre on a growing ch ain is transferred from a growing macromolecule - P• - or oligomer to another molecule (transfer agent XR) or to another site on the same molecule.

$$P\bullet + XR \rightarrow PX + R\bullet$$

This transfer involves termination of the initially growing chain to the completed mac-romolecule PX, where X denotes one end-group of the macromolecule. The example shows that the growing macromolecule as well as the transfer agent are consumed during this process. However, there are also chain transfer reactions that generate new chain carriers and new chain transfer agents at the same time with significant consequences for the distribution of the (average) molecular weight distribution, the dispersity Đ and the (average) degree of polymerization of the product. These chain transfer reactions are called degenerative chain transfer reactions and are observed, for example in RAFT-, ITRP-, or TERP- processes. RAFT = reversible addition-fragmen-tation chain transfer polymerization; ITRP = iodine-transfer polymerization; TERP = telluride-mediated polymerization. These polymerization techniques belong to the class of reversible deactivation radical polymerizations (RDRP) that show some char-acteristics of a living polymerization, however, they must not be addressed as living polymerizations because true living polymerizations are characterized by the absence of *any* termination reaction.

In this sense, a chain-transfer agent RX is a substance that is able to react with a chain carrier by a reaction in which the original chain P• is deactivated and a new chain carrier R• is generated.

Anionic Addition Polymerization

Anionic polymerization is a form of chain-growth polymerization that encompasses the polymerization of vinyl monomers with strong electronegative groups. This type of polymerization is often used to produce synthetic polydiene rubbers, solution styrene-butadiene rubbers (SBR), and thermoplastic styrenic elastomers.

All monomers with (strong) electronegative substituents polymerize readily in the presence of carbanions. Some electron-withdrawing substituents that stabilize the negative charge through charge delocalization and hence permit anionic polymerization include -CN, -COOR, $-C_6H_5$, and $-CH=CH_2$, to name only a few. Therefore, monomers such as styrenes, dienes, acrylates and methacrylates, aldehydes, epoxides, acrylonitriles and cyanoacrylates readily undergo anionic polymerization reactions.

The electron donors (or initiators) are either electron transfer agents or strong anions. The transfer of an electron from a donor molecule to the vinyl monomer leads to the formation of an anion radical, the so-called carbanion:

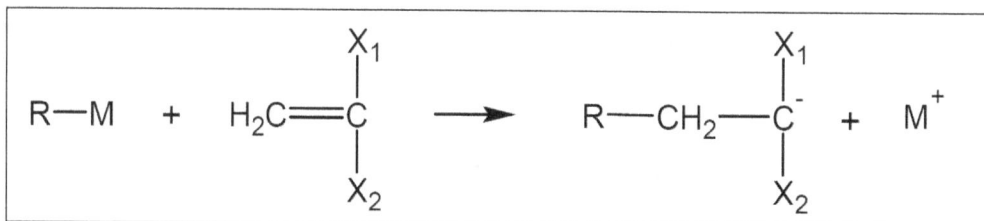

$$R-M \ + \ H_2C=\underset{X_2}{\overset{X_1}{C}} \ \longrightarrow \ R-CH_2-\underset{X_2}{\overset{X_1}{C^-}} \ + \ M^+$$

Typical electron donors (Lewis bases or nucleophiles) are alkali metals, such as lithium or sodium. Other strong nucleophilic initiators include covalent or ionic metal amides, alkoxides, hydroxides, amines, phosphines, cyanides, and organometallic compounds such as alkyl lithium compounds and Grignard reagents. The initiation proceeds by addition of a neutral (B:) or negative (B:⁻) nucleophile to the monomer.

The kinetics of an anionic polymerization consists of initiation, polymerization and termination. For example, the initiation and polymerization of styrene with potassium amide proceeds as follows:

$$KNH_2 \Leftrightarrow K^+ + NH_2^-$$

$$NH_2^- + M \rightarrow NH_2M^-$$

$$NH_2M_n^- + M \rightarrow NH_2M_{n+1}^-$$

$$NH_2M_n^- + NH_3 \rightarrow NH_2M_nH + NH_2^-$$

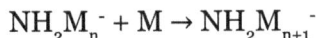

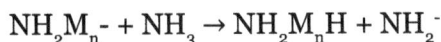

The "Gegen" ion, K^+, can be omitted from the scheme above, because it is dissolved ("free") in a media of comparatively high dielectric constant.

In carefully controlled systems (pure reactants and inert solvents), an anionic polymerization does not undergo termination reactions. Hence, the chains will remain active indefinitely unless there is deliberate termination or chain transfer. This has two important consequences:

1. The number average molecular weight, MW_n, of the polymer can be calculated from the amount of initiator and amount of consumed monomer, because the degree of polymerization is the ratio of the moles of monomer consumed to the moles of the initiator added: $MW_n = MW_0 [M_0] / [I]$, where MW_0 is the molecular weight of the repeat unit and $[M_0]$ and $[I]$ the (initial) concentrations of the monomer and the initiator.

2. Since all chains are initiated at roughly the same time, the polymer synthesis can be done in a controlled manner. In fact, it is the only one that leads to well defined and nearly mono-disperse molecular weight distribution (Poisson distribution) and structural and compositional uniformity.

This type of polymerization is called living polymerization. Anionic polymerization can also be used to functionalize polymers. The end-groups are usually added at the end of the polymerization. End-groups that have been used in the functionalization include -OH, -SH, -NH$_2$, COCH$_3$, -COOH, and epoxides, to name only a few.

Reversible-deactivation Radical Polymerization

Chain polymerization, propagated by radicals that are deactivated reversibly, bringing them into active/dormant equilibria of which there might be more than one.

Reversible deactivation radical polymerizations are members of the class of reversible deactivation polymerizations which exhibit much of the character of living polymerizations, but cannot be categorized as such as they are not without chain transfer or chain termination reactions. Several different names have been used in literature, which are:

*   Living radical polymerization.
*   Living free radical polymerization.
*   Controlled/"living" radical polymerization.

- Controlled radical polymerization.

- Reversible deactivation radical polymerization.

Though the term "living" radical polymerization was used in early days, it has been discouraged by IUPAC, because radical polymerization cannot be a truly living process due to unavoidable termination reactions between two radicals. The commonly used term controlled radical polymerization is permitted, but reversible-deactivated radical polymerization or controlled reversible-deactivation radical polymerization (RDRP) is recommended.

RDRP – sometimes misleadingly called 'free' radical polymerization – is one of the most widely used polymerization processes since it can be applied:

- To a great variety of monomers.

- It can be carried out in the presence of certain functional groups.

- The technique is rather simple and easy to control.

- The reaction conditions can vary from bulk over solution, emulsion, miniemulsion to suspension.

- It is relatively inexpensive compared with competitive techniques.

The steady-state concentration of the growing polymer chains is 10^{-7} M by order of magnitude, and the average life time of an individual polymer radical before termination is about 5–10 s. A drawback of the conventional radical polymerization is the limited control of chain architecture, molecular weight distribution, and composition. In the late 20th century it was observed that when certain components were added to systems polymerizing by a chain mechanism they are able to react reversibly with the (radical) chain carriers, putting them temporarily into a 'dormant' state. This had the effect of prolonging the lifetime of the growing polymer chains to values comparable with the duration of the experiment. At any instant most of the radicals are in the inactive (dormant) state, however, they are not irreversibly terminated ('dead'). Only a small fraction of them are active (growing), yet with a fast rate of interconversion of active and dormant forms, faster than the growth rate, the same probability of growth is ensured for all chains, i.e., on average, all chains are growing at the same rate. Consequently, rather than a most probable distribution, the molecular masses (degrees of polymerization) assume a much narrower Poisson distribution, and a lower dispersity prevails.

IUPAC also recognizes the alternative name, 'controlled reversible-deactivation radical polymerization' as acceptable, "provided the controlled context is specified, which in this instance comprises molecular mass and molecular mass distribution." These types of radical polymerizations are not necessarily 'living' polymerizations, since chain termination reactions are not precluded".

The adjective 'controlled' indicates that a certain kinetic feature of a polymerization or structural aspect of the polymer molecules formed is controlled (or both). The expression 'controlled polymerization' is sometimes used to describe a radical or ionic polymerization in which reversible-deactivation of the chain carriers is an essential component of the mechanism and interrupts the propagation that secures control of one or more kinetic features of the polymerization or one or more structural aspects of the macromolecules formed, or both. The expression 'controlled radical polymerization' is sometimes used to describe a radical polymerization that is conducted in the presence of agents that lead to e.g. atom-transfer radical polymerization (ATRP), nitroxide-(aminoxyl) mediated polymerization (NMP), or reversible-addition-fragmentation chain transfer (RAFT) polymerization. All these and further controlled polymerizations are included in the class of reversible-deactivation radical polymerizations. Whenever the adjective 'controlled' is used in this context the particular kinetic or the structural features that are controlled have to be specified.

Reversible-deactivation Polymerization

There is a mode of polymerization referred to as reversible-deactivation polymerization which is distinct from living polymerization, despite some common features. Living polymerization requires a complete absence of termination reactions, whereas reversible-deactivation polymerization may contain a similar fraction of termination as conventional polymerization with the same concentration of active species. Some important aspects of these are compared in the table:

Comparison of radical polymerization processes			
Property	Standard radical polymerization	Living polymerization	Reversible-deactivation polymerization
Concn. of initiating species	Falls off only slowly	Falls off very rapidly	Falls off very rapidly
Concn. of chain carriers (Number of growing chains)	Instantaneous steady state (Bodenstein approximation applies) decreasing throughout reaction	Constant throughout reaction	Constant throughout reaction
Lifetime of growing chains	$\sim 10^{-3}$ s	Same as reaction duration	Same as reaction duration
Main form of termination	Radical combination or radical disproportionation	Termination reactions are precluded	Termination reactions are not precluded
Molar mass distribution	Broad range (Đ >=1.5), Schulz-Zimm distribution	Narrow range(Đ <1.5), Poisson distribution	Narrow range(Đ <1.5), Poisson distribution
Dormant states	None	Rare	Predominant

Common Features

As the name suggests, the prerequisite of a successful RDRP is fast and reversible activation/deactivation of propagating chains. There are three types of RDRP; namely deactivation by catalyzed reversible coupling, deactivation by spontaneous reversible coupling and deactivation by degenerative transfer (DT). A mixture of different mechanisms is possible; e.g. a transition metal mediated RDRP could switch among ATRP, OMRP and DT mechanisms depending on the reaction conditions and reagents used.

In any RDRP processes, the radicals can propagate with the rate coefficient k_p by addition of a few monomer units before the deactivation reaction occurs to regenerate the dormant species. Concurrently, two radicals may react with each other to form dead chains with the rate coefficient k_t. The rates of propagation and termination between two radicals are not influenced by the mechanism of deactivation or the catalyst used in the system. Thus it is possible to estimate how fast a RDRP can be conducted with preserved chain end functionality?

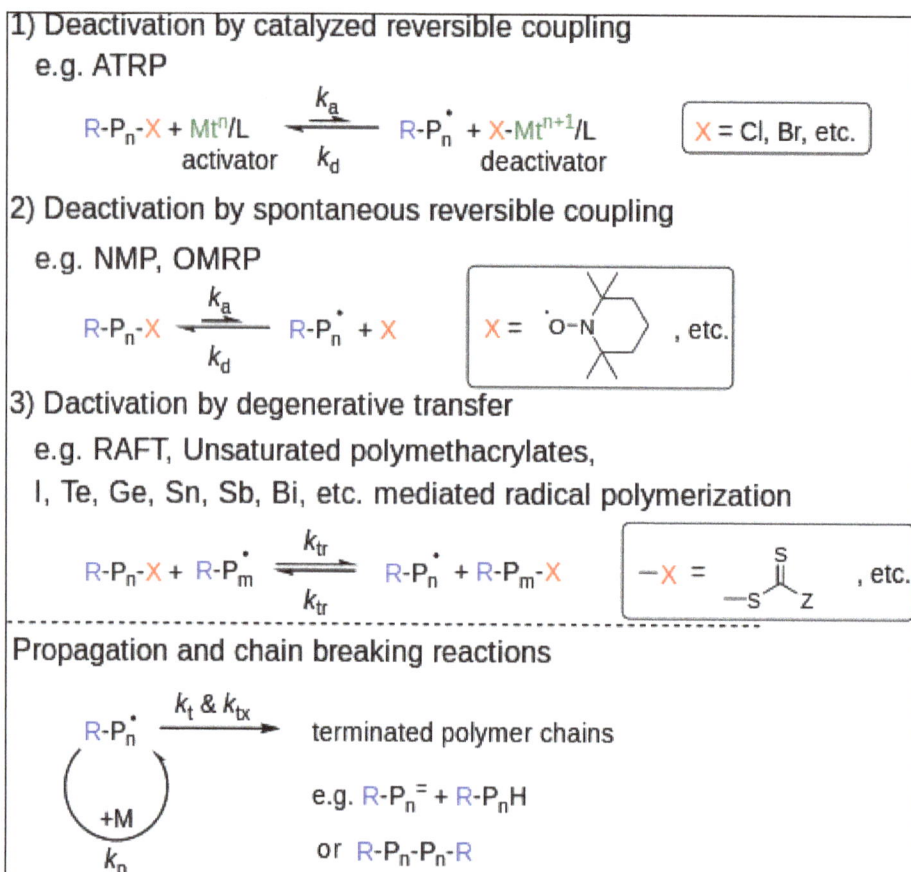

In addition, other chain breaking reactions such as irreversible chain transfer/termination reactions of the propagating radicals with solvent, monomer, polymer, catalyst, additives, etc. would introduce additional loss of chain end functionality (CEF). The

overall rate coefficient of chain breaking reactions besides the direct termination between two radicals is represented as k_{tx}.

In all RDRP methods, the theoretical number average molecular weight of obtained polymers, M_n, can be defined by following equation:

$$M_n = M_m \times \frac{[M]_0 - [M]_t}{[R\text{-}X]_0}$$

where M_m is the molecular weight of monomer; $[M]_0$ and $[M]_t$ are the monomer concentrations at time 0 and time t; $[R\text{-}X]_0$ is the initial concentration of the initiator.

Besides the designed molecular weight, a well controlled RDRP should give polymers with narrow molecular distributions, which can be quantified by M_w/M_n values, and well preserved chain end functionalities.

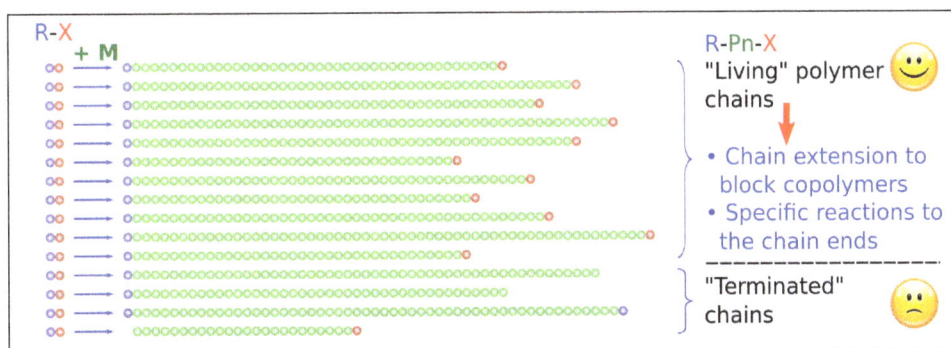

A well controlled RDRP process requires: 1) the reversible deactivation process should be sufficiently fast; 2) the chain breaking reactions which cause the loss of chain end functionalities should be limited; 3) properly maintained radical concentration; 4) the initiator should have proper activity.

Atom Transfer Radical Polymerization (ATRP)

The initiator of the polymerization is usually an organohalogenid and the dormant state is achieved in a metal complex of a transition metal ('radical buffer'). This method is very versatile but requires unconventional initiator systems that are sometimes poorly compatible with the polymerization media.

Nitroxide-mediated Polymerization (NMP)

Given certain conditions a homolytic splitting of the C-O bond in alkoxylamines can occur and a stable 2-centre 3 electron N-O radical can be formed that is able to initiate a polymerization reaction. The preconditions for an alkoxylamine suitable to initiate a polymerization are bulky, sterically obstructive substituents on the secondary amine, and the substituent on the oxygen should be able to form a stable radical, e.g. benzyl.

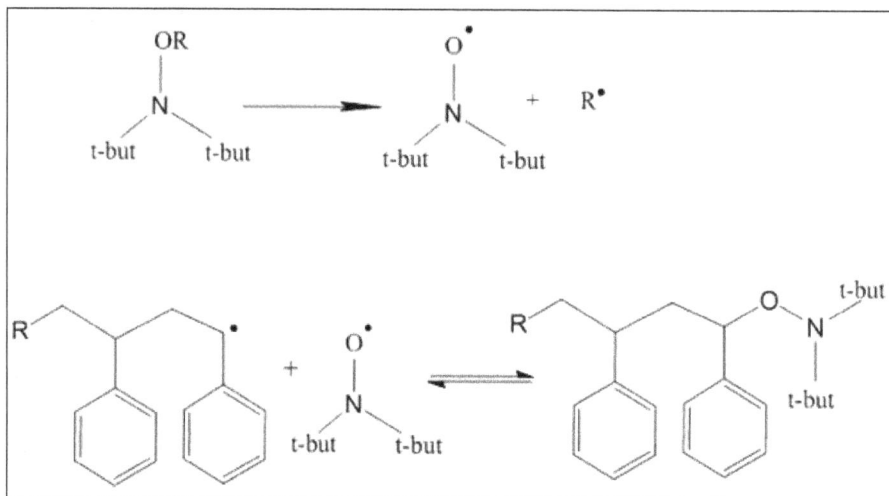

Example of a reversible deactivation reaction.

Reversible Addition-fragmentation Chain Transfer (RAFT)

RAFT is one of the most versatile and convenient techniques in this context. The most common RAFT-processes are carried out in the presence of thiocarbonylthio compounds that act as radical buffers. While in ATRP and NMP reversible deactivation of propagating radical-radical reactions takes place and the dormant structures are a halo-compound in ATRP and the alkoxyamine in NMP, both being a sink for radicals and source at the same time and described by the corresponding equilibria. RAFT on the contrary, is controlled by chain-transfer reactions that are in a deactivation-activation equilibrium. Since no radicals are generated or destroyed an external source of radicals is necessary for initiation and maintenance of the propagation reaction.

Initiation step of a RAFT polymerization

$$I \rightarrow I \xrightarrow{M \; M} \rightarrow P_n^{\cdot}$$

Reversible chain transfer:

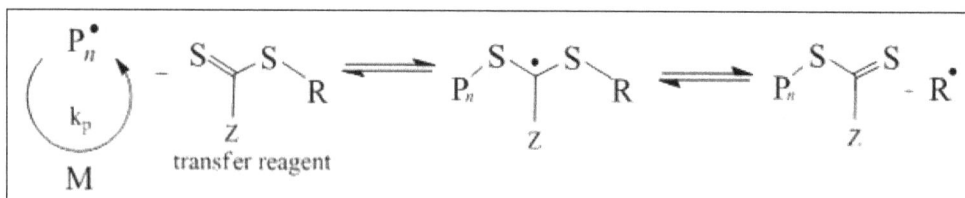

Reinitiation step:

$$R^{\cdot} \xrightarrow{M} RM^{\cdot} \xrightarrow{M \; M} \rightarrow P_m^{\cdot}$$

Chain equilibration step:

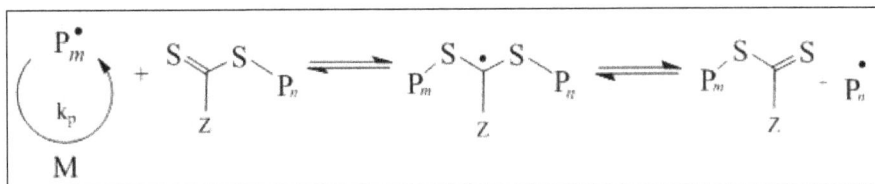

Termination step:

$$P_m^{\cdot} + P_n^{\cdot} \rightarrow P_m P_n$$

Catalytic Chain Transfer and Cobalt Mediated Radical Polymerization

Although not a strictly living form of polymerization catalytic chain transfer polymerization must be mentioned as it figures significantly in the development of later forms of living free radical polymerization. Discovered in the late 1970s in the USSR it was found that cobalt porphyrins were able to reduce the molecular weight during polymerization of methacrylates. Later investigations showed that the cobalt glyoxime complexes were as effective as the porphyrin catalysts and also less oxygen sensitive. Due to their lower oxygen sensitivity these catalysts have been investigated much more thoroughly than the porphyrin catalysts.

The major products of catalytic chain transfer polymerization are vinyl-terminated polymer chains. One of the major drawbacks of the process is that catalytic chain transfer polymerization does not produce macromonomers but instead produces addition fragmentation agents. When a growing polymer chain reacts with the addition fragmentation agent the radical end-group attacks the vinyl bond and forms a bond. However, the resulting product is so hindered that the species undergoes fragmentation, leading eventually to telechelic species.

These addition fragmentation chain transfer agents do form graft copolymers with styrenic and acrylate species however they do so by first forming block copolymers and then incorporating these block copolymers into the main polymer backbone.

While high yields of macromonomers are possible with methacrylate monomers, low yields are obtained when using catalytic chain transfer agents during the polymerization of acrylate and stryenic monomers. This has been seen to be due to the interaction of the radical centre with the catalyst during these polymerization reactions.

The reversible reaction of the cobalt macrocycle with the growing radical is known as cobalt carbon bonding and in some cases leads to living polymerization reactions.

Iniferter Polymerization

An iniferter is a chemical compound that simultaneously acts as initiator, transfer agent, and terminator (hence the name ini-fer-ter) in controlled free radical iniferter polymerizations, the most common is the dithiocarbamate type.

Iodine-transfer Polymerization (ITP)

Iodine-transfer polymerization (ITP, also called ITRP), developed by Tatemoto and coworkers in the 1970s gives relatively low polydispersities for fluoroolefin polymers. While it has received relatively little academic attention, this chemistry has served as the basis for several industrial patents and products and may be the most commercially successful form of living free radical polymerization. It has primarily been used to incorporate iodine cure sites into fluoroelastomers.

The mechanism of ITP involves thermal decomposition of the radical initiator (typically persulfate), generating the initiating radical In•. This radical adds to the monomer M to form the species P_1•, which can propagate to P_m•. By exchange of iodine from the transfer agent R-I to the propagating radical P_m• a new radical R• is formed and P_m• becomes dormant. This species can propagate with monomer M to P_n•. During the polymerization exchange between the different polymer chains and the transfer agent occurs, which is typical for a degenerative transfer process.

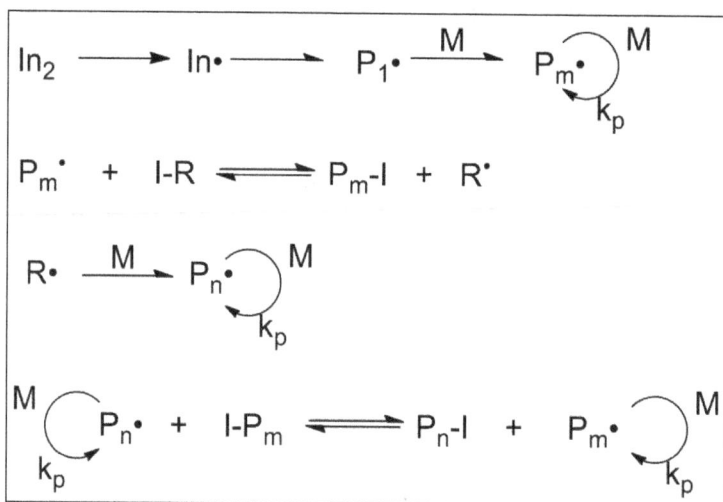

Typically, iodine transfer polymerization uses a mono- or diiodo-perfluoroalkane as the initial chain transfer agent. This fluoroalkane may be partially substituted with hydrogen or chlorine. The energy of the iodine-perfluoroalkane bond is low and, in contrast to iodo-hydrocarbon bonds, its polarization small. Therefore, the iodine is easily abstracted in the presence of free radicals. Upon encountering an iodoperfluoroalkane, a growing poly(fluoroolefin) chain will abstract the iodine and terminate, leaving the now-created perfluoroalkyl radical to add further monomer. But the iodine-terminated poly(fluoroolefin) itself acts as a chain transfer agent. As in RAFT processes, as long as the rate of initiation is kept low, the net result is the formation of a monodisperse molecular weight distribution.

Use of conventional hydrocarbon monomers with iodoperfluoroalkane chain transfer agents has been described. The resulting molecular weight distributions have not been narrow since the energetics of an iodine-hydrocarbon bond are considerably different from that of an iodine-fluorocarbon bond and abstraction of the iodine from the terminated polymer difficult. The use of hydrocarbon iodides has also been described, but again the resulting molecular weight distributions were not narrow.

Preparation of block copolymers by iodine-transfer polymerization was also described by Tatemoto and coworkers in the 1970s.

Although use of living free radical processes in emulsion polymerization has been characterized as difficult, all examples of iodine-transfer polymerization have involved emulsion polymerization. Extremely high molecular weights have been claimed.

Listed below are some other less described but to some extent increasingly important living radical polymerization techniques.

Selenium-centered Radical-mediated Polymerization

Diphenyl diselenide and several benzylic selenides have been explored by Kwon *et al.* as photoiniferters in polymerization of styrene and methyl methacrylate. Their mechanism of control over polymerization is proposed to be similar to the dithiuram disulfide iniferters. However, their low transfer constants allow them to be used for block copolymer synthesis but give limited control over the molecular weight distribution.

Telluride-mediated Polymerization (TERP)

Telluride-mediated polymerization or TERP first appeared to mainly operate under a reversible chain transfer mechanism by homolytic substitution under thermal initiation. However, in a kinetic study it was found that TERP predominantly proceeds by degenerative transfer rather than 'dissociation combination'.

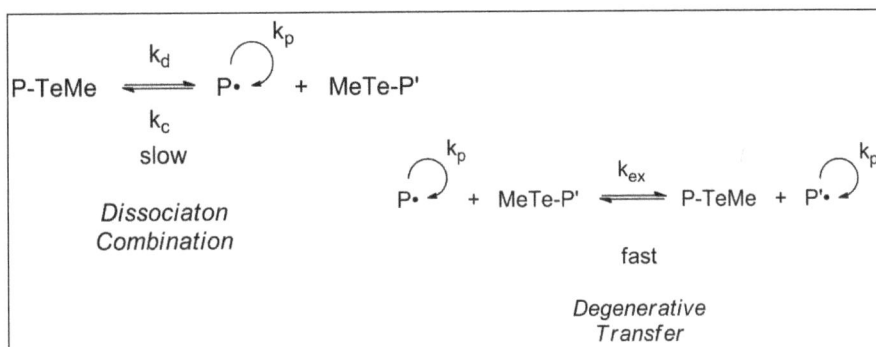

Alkyl tellurides of the structure Z-X-R, were Z=methyl and R= a good free radical leaving group, give the better control for a wide range of monomers, phenyl tellurides (Z=phenyl) giving poor control. Polymerization of methyl methacrylates are only controlled by ditellurides. The importance of X to chain transfer increases in the series O<S<Se<Te, makes alkyl tellurides effective in mediating control under thermally initiated conditions and the alkyl selenides and sulfides effective only under photoinitiated polymerization.

Stibine-mediated Polymerization

More recently Yamago *et al.* reported stibine-mediated polymerization, using an organostibine transfer agent with the general structure Z(Z')-Sb-R (where Z= activating group and R= free radical leaving group). A wide range of monomers (styrenics, (meth) acrylics and vinylics) can be controlled, giving narrow molecular weight distributions and predictable molecular weights under thermally initiated conditions. Yamago has also published a patent indicating that bismuth alkyls can also control radical polymerizations via a similar mechanism.

Copper Mediated Polymerization

More reversible-deactivation radical polymerizations are known to be catalysed by copper.

Atom Transfer Radical Polymerization

Atom transfer radical polymerization (ATRP) is an example of a reversible-deactivation radical polymerization. Like its counterpart, ATRA, or atom transfer radical addition, ATRP is a means of forming a carbon-carbon bond with a transition metal catalyst. The polymerization from this method is called atom transfer radical addition polymerization (ATRAP). As the name implies, the atom transfer step is crucial in the reaction responsible for uniform polymer chain growth. ATRP (or transition metal-mediated living radical polymerization) was independently discovered by Mitsuo Sawamoto and by Krzysztof Matyjaszewski in 1995.

The following scheme presents a typical ATRP reaction:

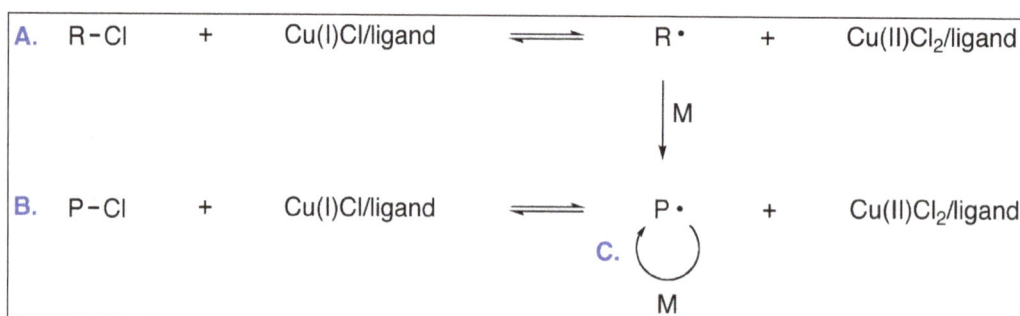

A.	R−Cl	+	Cu(I)Cl/ligand	⇌	R•	+	Cu(II)Cl$_2$/ligand
						M	
B.	P−Cl	+	Cu(I)Cl/ligand	⇌	P•	+	Cu(II)Cl$_2$/ligand
					C. ◯		
					M		

General ATRP reaction. A. Initiation. B. Equilibrium with dormant species. C. Propagation.

ATRP usually employs a transition metal complex as the catalyst with an alkyl halide as the initiator (R-X). Various transition metal complexes, namely those of Cu, Fe,

Ru, Ni, and Os, have been employed as catalysts for ATRP. In an ATRP process, the dormant species is activated by the transition metal complex to generate radicals via one electron transfer process. Simultaneously the transition metal is oxidized to higher oxidation state. This reversible process rapidly establishes an equilibrium that is predominately shifted to the side with very low radical concentrations. The number of polymer chains is determined by the number of initiators. Each growing chain has the same probability to propagate with monomers to form living/dormant polymer chains (R-P_n-X). As a result, polymers with similar molecular weights and narrow molecular weight distribution can be prepared.

ATRP reactions are very robust in that they are tolerant of many functional groups like allyl, amino, epoxy, hydroxy, and vinyl groups present in either the monomer or the initiator. ATRP methods are also advantageous due to the ease of preparation, commercially available and inexpensive catalysts (copper complexes), pyridine-based ligands, and initiators (alkyl halides).

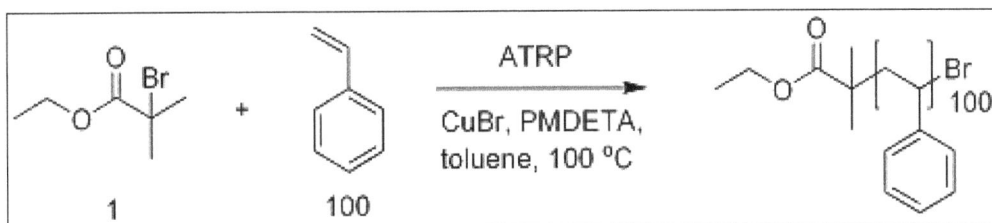

The ATRP with styrene. If all the styrene is reacted (the conversion is 100%) the polymer will have 100 units of styrene built into it. PMDETA stands for *N,N,N',N'',N''*-pentamethyldiethylenetriamine.

Components of Normal ATRP

There are five important variable components of atom transfer radical polymerizations. They are the monomer, initiator, catalyst, ligand, and solvent.

Monomer

Monomers typically used in ATRP are molecules with substituents that can stabilize the propagating radicals; for example, styrenes, (meth)acrylates, (meth)acrylamides, and acrylonitrile. ATRP is successful at leading to polymers of high number average molecular weight and low dispersity when the concentration of the propagating radical balances the rate of radical termination. Yet, the propagating rate is unique to each individual monomer. Therefore, it is important that the other components of the polymerization (initiator, catalyst, ligand, and solvent) are optimized in order for the concentration of the dormant species to be greater than that of the propagating radical while being low enough as to prevent slowing down or halting the reaction.

Initiator

The number of growing polymer chains is determined by the initiator. To ensure a low polydispersity and a controlled polymerization, the rate of initiation must be as fast or preferably faster than the rate of propagation. Ideally, all chains will be initiated in a very short period of time and will be propagated at the same rate. Initiators are typically chosen to be alkyl halides whose frameworks are similar to that of the propagating radical. Alkyl halides such as alkyl bromides are more reactive than alkyl chlorides. Both offer good molecular weight control. The shape or structure of the initiator influences polymer architecture. For example, initiators with multiple alkyl halide groups on a single core can lead to a star-like polymer shape. Furthermore, α-functionalized ATRP initiators can be used to synthesize hetero-telechelic polymers with a variety of chain-end groups.

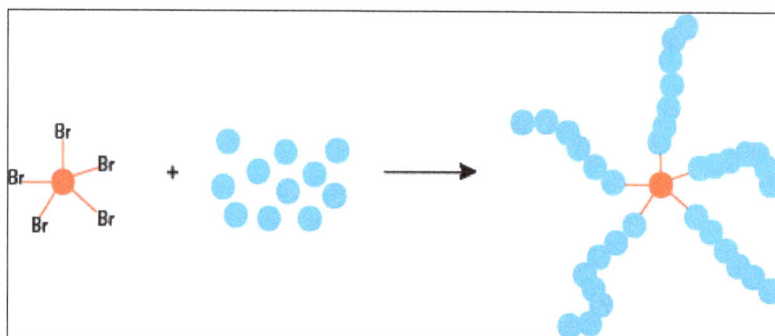

Illustration of a star initiator for ATRP.

Catalyst

The catalyst is the most important component of ATRP because it determines the equilibrium constant between the active and dormant species. This equilibrium determines the polymerization rate. An equilibrium constant that is too small may inhibit or slow the polymerization while an equilibrium constant that is too large leads to a wide distribution of chain lengths.

There are several requirements for the metal catalyst:

1. There needs to be two accessible oxidation states that are differentiated by one electron.

2. The metal center needs to have reasonable affinity for halogens.

3. The coordination sphere of the metal needs to be expandable when it is oxidized as to accommodate the halogen.

4. The transition metal catalyst should not lead to significant side reactions, such as irreversible coupling with the propagating radicals and catalytic radical termination.

The most studied catalysts are those that include copper, which has shown the most versatility with successful polymerizations for a wide selection of monomers.

Ligand

One of the most important aspects in an ATRP reaction is the choice of ligand which is used in combination with the traditionally copper halide catalyst to form the catalyst complex. The main function of the ligand is to solubilize the copper halide in whichever solvent is chosen and to adjust the redox potential of the copper. This changes the activity and dynamics of the halogen exchange reaction and subsequent activation and deactivation of the polymer chains during polymerization, therefore greatly affecting the kinetics of the reaction and the degree of control over the polymerization. Different ligands should be chosen based on the activity of the monomer and the choice of metal for the catalyst. As copper halides are primarily used as the catalyst, amine based ligands are most commonly chosen. Ligands with higher activities are being investigated as ways to potentially decrease the concentration of catalyst in the reaction since a more active catalyst complex would lead to a higher concentration of deactivator in the reaction. However, a too active catalyst can lead to a loss of control and increase the polydispersity of the resulting polymer.

Solvents

Toluene, 1,4-dioxane, xylene, anisole, DMF, DMSO, water, methanol, acetonitrile, or even the monomer itself (described as a bulk polymerization) are commonly used.

Kinetics of Normal ATRP

- Reactions in atom transfer radical polymerization:

  Initiation:

  $$R\text{-}X + Cu^I X/L \underset{k_{d,0}}{\overset{k_{a,0}}{\rightleftharpoons}} Cu^{II} X_2/L + R^{\cdot} \qquad K_{ATRP,0} = \frac{k_{a,0}}{k_{d,0}}$$

  $$R^{\cdot} + M \xrightarrow{k_{add}} R\text{-}P_1^{\cdot}$$

  $$2R^{\cdot} \xrightarrow{k_{t,0}} \begin{cases} R\text{-}R \\ \text{or} \\ R^= + RH \end{cases}$$

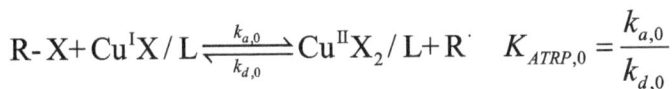

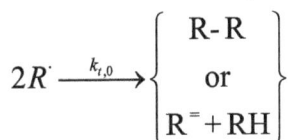

  Quasi-steady state:

- Other chain breaking reactions (k_{tx}) should also be considered.

$$\text{R-P}_n\text{-X} + \text{Cu}^I\text{X/L} \underset{k_d}{\overset{k_a}{\rightleftharpoons}} \text{Cu}^{II}\text{X}_2\text{/L} + \text{R-P}_n^{\bullet}$$

ATRP

activation/ deactivation

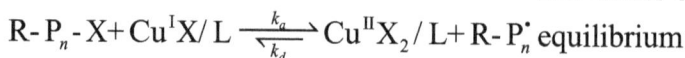

equilibrium

$$k_{ATRP} = \frac{k_a}{k_d}$$

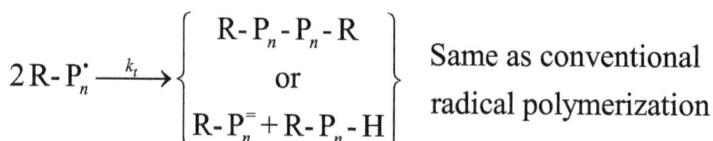

$$\text{R-P}_n^{\bullet} + \text{M} \xrightarrow{k_p} \text{R-P}_{n+1}^{\bullet}$$

$$2\,\text{R-P}_n^{\bullet} \xrightarrow{k_t} \left\{ \begin{array}{c} \text{R-P}_n\text{-P}_n\text{-R} \\ \text{or} \\ \text{R-P}_n^{=} + \text{R-P}_n\text{-H} \end{array} \right\}$$

Same as conventional radical polymerization

ATRP Equilibrium Constant

The radical concentration in normal ATRP can be calculated via the following equation:

$$[\text{R-P}_n^{\bullet}] = K_{ATRP} \cdot [\text{R-P}_n\text{-X}] \cdot \frac{[\text{Cu}^I\text{X/L}]}{[\text{Cu}^{II}\text{X}2\text{/L}]}$$

It is important to know the K_{ATRP} value to adjust the radical concentration. The K_{ATRP} value depends on the homo-cleavage energy of the alkyl halide and the redox potential of the Cu catalyst with different ligands. Given two alkyl halides (R¹-X and R²-X) and two ligands (L¹ and L²), there will be four combinations between different alkyl halides and ligands. Let K^{ij}_{ATRP} refer to the K_{ATRP} value for R^i-X and L^j. If we know three of these four combinations, the fourth one can be calculated as:

$$K^{22}_{ATRP} = \frac{K^{12}_{ATRP} \times K^{21}_{ATRP}}{K^{11}_{ATRP}}$$

The K_{ATRP} values for different alkyl halides and different Cu catalysts can be found in literature.

Solvents have significant effects on the K_{ATRP} values. The K_{ATRP} value increases dramatically with the polarity of the solvent for the same alkyl halide and the same Cu catalyst. The polymerization must take place in solvent/monomer mixture, which changes to solvent/monomer/polymer mixture gradually. The K_{ATRP} values could change 10000 times by switching the reaction medium from pure methyl acrylate to pure dimethyl sulfoxide.

Activation and Deactivation Rate Coefficients

Deactivation rate coefficient, k_d, values must be sufficiently large to obtain low dispersity. The direct measurement of k_d is difficult though not impossible. In most cases, k_d

may be calculated from known K_{ATRP} and k_a. Cu complexes providing very low k_d values are not recommended for use in ATRP reactions.

Retention of Chain End Functionality

$$\sum[X] = Constant$$

$$\underbrace{[R\text{-}X]_0 - [R\text{-}X]_t - [R\text{-}P_n\text{-}X]_t}_{\substack{\text{Loss of chain} \\ \text{end functionality}}} = \underbrace{([Cu^I X/L]_t + [Cu^{II} X_2/L]_t) - ([Cu^I X/L]_0 + 2[Cu^{II} X_2/L]_0)}_{\text{Change in } [Cu^I X/L] \text{ and } [Cu^{II} X2/L]} + \underbrace{[RA\text{-}X]_t}_{\substack{\text{X transfer in} \\ \text{activator} \\ \text{regeneration}}}$$

High level retention of chain end functionality is typically desired. However, the determination of the loss of chain end functionality based on ^{1}H NMR and mass spectroscopy methods cannot provide precise values. As a result, it is difficult to identify the contributions of different chain breaking reactions in ATRP. One simple rule in ATRP comprises the principle of halogen conservation. Halogen conservation means the total amount of halogen in the reaction systems must remain as a constant. From this rule, the level of retention of chain end functionality can be precisely determined in many cases. The precise determination of the loss of chain end functionality enabled further investigation of the chain breaking reactions in ATRP.

Advantages of ATRP

ATRP enables the polymerization of a wide variety of monomers with different chemical functionalities, proving to be more tolerant of these functionalities than ionic polymerizations. It provides increased control of molecular weight, molecular architecture and polymer composition while maintaining a low polydispersity (1.05-1.2). The halogen remaining at the end of the polymer chain after polymerization allows for facile post-polymerization chain-end modification into different reactive functional groups. The use of multi-functional initiators facilitates the synthesis of lower-arm star polymers and telechelic polymers.

Disadvantages of ATRP

The most significant drawback of ATRP is the high concentrations of catalyst required for the reaction. This catalyst standardly consists of a copper halide and an amine-based ligand. The removal of the copper from the polymer after polymerization is often tedious and expensive, limiting ATRP's use in the commercial sector. However, researchers are currently developing methods which would limit the necessity of the catalyst concentration to ppm. ATRP is also a traditionally air-sensitive reaction normally requiring freeze-pump thaw cycles. However, techniques such as Activator Generated by Electron Transfer (AGET) ATRP provide potential alternatives which are not air-sensitive. A final disadvantage is the difficulty of conducting ATRP in aqueous media.

Different ATRP Methods

Activator Regeneration ATRP Methods

In a normal ATRP, the concentration of radicals is determined by the K_{ATRP} value, concentration of dormant species, and the $[Cu^I]/[Cu^{II}]$ ratio. In principle, the total amount of Cu catalyst should not influence polymerization kinetics. However, the loss of chain end functionality slowly but irreversibly converts Cu^I to Cu^{II}. Thus initial $[Cu^I]/[I]$ ratios are typically 0.1 to 1. When very low concentrations of catalysts are used, usually at the ppm level, activator regeneration processes are generally required to compensate the loss of CEF and regenerate a sufficient amount of Cu^I to continue the polymerization. Several activator regeneration ATRP methods were developed, namely ICAR ATRP, ARGET ATRP, SARA ATRP, eATRP, and photoinduced ATRP. The activator regeneration process is introduced to compensate the loss of chain end functionality, thus the cumulative amount of activator regeneration should roughly equal the total amount of the loss of chain end functionality.

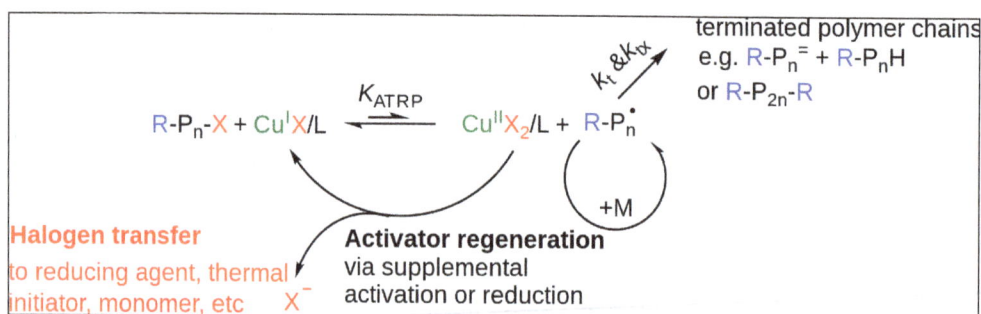

Activator regeneration Atom Transfer Radical Polymerization.

ICAR ATRP

Initiators for continuous activator regeneration (ICAR) is a technique that uses conventional radical initiators to continuously regenerate the activator, lowering its required concentration from thousands of ppm to <100 ppm; making it an industrially relevant technique.

ARGET ATRP

Activators regenerated by electron transfer (ARGET) employs non-radical forming reducing agents for regeneration of Cu^I. A good reducing agent (e.g. hydrazine, phenols, sugars, ascorbic acid) should only react with Cu^{II} and not with radicals or other reagents in the reaction mixture.

SARA ATRP

A typical SARA ATRP employs Cu^0 as both supplemental activator and reducing agent (SARA). Cu^0 can activate alkyl halide directly but slowly. Cu^0 can also reduce Cu^{II} to Cu^I.

Both processes help to regenerate Cu^I activator. Other zerovalent metals, such as Mg, Zn, and Fe, have also been employed for Cu-based SARA ATRP.

eATRP

In *eATRP* the activator Cu^I is regenerated via electrochemical process. The development of *eATRP* enables precise control of the reduction process and external regulation of the polymerization. In an *eATRP* process, the redox reaction involves two electrodes. The Cu^{II} species is reduced to Cu^I at the cathode. The anode compartment is typically separated from the polymerization environment by a glass frit and a conductive gel. Alternatively, a sacrificial aluminum counter electrode can be used, which is directly immersed in the reaction mixture.

Photoinduced ATRP

The direct photo reduction of transition metal catalysts in ATRP and/or photo assistant activation of alkyl halide is particularly interesting because such a procedure will allow performing of ATRP with ppm level of catalysts without any other additives.

Other ATRP Methods

Reverse ATRP

In reverse ATRP, the catalyst is added in its higher oxidation state. Chains are activated by conventional radical initiators (e.g. AIBN) and deactivated by the transition metal. The source of transferrable halogen is the copper salt, so this must be present in concentrations comparable to the transition metal.

SR&NI ATRP

A mixture of radical initiator and active (lower oxidation state) catalyst allows for the creation of block copolymers (contaminated with homopolymer) which is impossible using standard reverse ATRP. This is called SR&NI (simultaneous reverse and normal initiation ATRP).

AGET ATRP

Activators generated by electron transfer uses a reducing agent unable to initiate new chains (instead of organic radicals) as regenerator for the low-valent metal. Examples are metallic copper, tin(II), ascorbic acid, or triethylamine. It allows for lower concentrations of transition metals, and may also be possible in aqueous or dispersed media.

Hybrid and Bimetallic Systems

This technique uses a variety of different metals/oxidation states, possibly on solid supports, to act as activators/deactivators, possibly with reduced toxicity or sensitivity. Iron salts can, for example, efficiently activate alkyl halides but requires an efficient Cu(II) deactivator which can be present in much lower concentrations (3–5 mol%).

Metal-free ATRP

Trace metal catalyst remaining in the final product has limited the application of ATRP in biomedical and electronic fields. In 2014, Craig Hawker and coworkers developed a new catalysis system involving photoredox reaction of 10-phenothiazine. The metal-free ATRP has been demonstrated to be capable of controlled polymerization of methacrylates. This technique was later expanded to polymerization of acrylonitrile by Matyjaszewski et al.

Mechano/sono-ATRP

Mechano/sono-ATRP uses mechanical forces, typically ultrasonic agitation, as an external stimulus to induce the (re)generation of activators in ATRP. Esser-Kahn, et al. demonstrated the first example of mechanoATRP using the piezoelectricity of barium titanate to reduce Cu(II) species. Matyjaszewski, et al. later improved the technique by using nanometer-sized and/or surface-functionalized barium titanate or zinc oxide particles, achieving superior rate and control of polymerization, as well as temporal control, with ppm-level of copper catalysts. In addition to peizoelectric particles, water and carbonates were found to mediate mechano/sono-ATRP. Mechochemically homolyzed water molecules undergoes radical addition to monomers, which in turn reduces Cu(II) species. Mechanically unstable Cu(II)-carbonate complexes formed in the presence to insoluble carbonates, which oxidizes dimethylsulfoxide, the solvent molecules, to generate Cu(I) species and carbon dioxide.

Polymers Synthesized through ATRP

- Polystyrene

- Poly (methyl methacrylate)

- Polyacrylamide

Reversible Addition-fragmentation Chain-transfer Polymerization

Reversible addition-fragmentation chain transfer or RAFT polymerization is one of several kinds of reversible-deactivation radical polymerization. It makes use of a chain transfer agent in the form of a thiocarbonylthio compound to afford control over the generated molecular weight and polydispersity during a free-radical polymerization. Discovered at the Commonwealth Scientific and Industrial Research Organisation (CSIRO) of Australia in 1998, RAFT polymerization is one of several living or controlled radical polymerization techniques, others being atom transfer radical polymerization (ATRP) and nitroxide-mediated polymerization (NMP), etc. RAFT polymerization uses thiocarbonylthio compounds, such as dithioesters, thiocarbamates, and xanthates, to mediate the polymerization via a reversible chain-transfer process. As with other controlled radical polymerization techniques, RAFT polymerizations can be performed with conditions to favor low dispersity (molecular weight distribution) and a pre-chosen molecular weight. RAFT polymerization can be used to design polymers of complex architectures, such as linear block copolymers, comb-like, star, brush polymers, dendrimers and cross-linked networks.

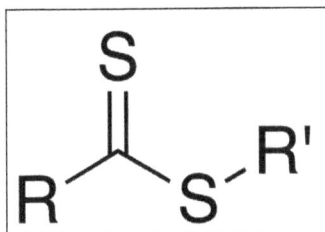

Structure of a thiocarbonylthio.

The addition-fragmentation chain transfer process was first reported in the early 1970s. However, the technique was irreversible, so the transfer reagents could not be used to control radical polymerization at this time. For the first few years addition-fragmentation chain transfer was used to help synthesize end-functionalized polymers.

Scientists began to realize the potential of RAFT in controlled radical polymerization in the 1980s. Macromonomers were known as reversible chain transfer agents during this time, but had limited applications on controlled radical polymerization.

In 1995, a key step in the "degenerate" reversible chain transfer step for chain equilibration was brought to attention. The essential feature is that the product of chain transfer is also a chain transfer agent with similar activity to the precursor transfer agent.

RAFT polymerization today is mainly carried out by thiocarbonylthio chain transfer agents. It was first reported by Rizzardo *et al.* in 1998. RAFT is one of the most versatile methods of controlled radical polymerization because it is tolerant of a very wide range of functionality in the monomer and solvent, including aqueous solutions. RAFT polymerization has also been effectively carried out over a wide temperature range.

Important Components of RAFT

Typically, a RAFT polymerization system consists of:

- A radical source (e.g. thermochemical initiator or the interaction of gamma radiation with some reagent).

- Monomer.

- RAFT agent.

- Solvent (not strictly required if the monomer is a liquid).

Two examples of RAFT agents.

A temperature is chosen such that (a) chain growth occurs at an appropriate rate, (b) the chemical initiator (radical source) delivers radicals at an appropriate rate and (c) the central RAFT equilibrium favors the active rather than dormant state to an acceptable extent.

RAFT polymerization can be performed by adding a chosen quantity of an appropriate RAFT agent to a conventional free radical polymerization. Usually the same monomers, initiators, solvents and temperatures can be used.

Radical initiators such as azobisisobutyronitrile (AIBN) and 4,4'-azobis(4-cyanovaleric acid) (ACVA), also called 4,4'-azobis(4-cyanopentanoic acid), are widely used as the initiator in RAFT.

Figure provides a visual description of RAFT polymerizations of poly(methyl methacrylate) and polyacrylic acid using AIBN as the initiator and two RAFT agents.

RAFT polymerization is known for its compatibility with a wide range of monomers compared to other controlled radical polymerizations. These monomers include (meth) acrylates, (meth)acrylamides, acrylonitrile, styrene and derivatives, butadiene, vinyl acetate and N-vinylpyrrolidone. The process is also suitable for use under a wide range of reaction parameters such as temperature or the level of impurities, as compared to NMP or ATRP.

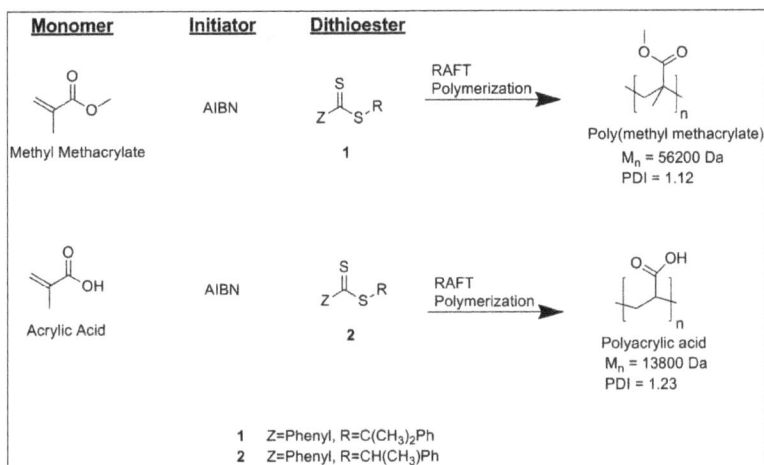

Examples of the major reagents and products in two RAFT polymerizations.

The Z and R group of a RAFT agent must be chosen according to a number of considerations. The Z group primarily affects the stability of the S=C bond and the stability of the adduct radical (Polymer-S-C•(Z)-S-Polymer, see section on Mechanism). These in turn affect the position of and rates of the elementary reactions in the pre- and main-equilibrium. The R group must be able to stabilize a radical such that the right hand side of the pre-equilibrium is favored, but unstable enough that it can reinitiate growth of a new polymer chain. As such, a RAFT agent must be designed with consideration of the monomer and temperature, since both these parameters also strongly influence the kinetics and thermodynamics of the RAFT equilibria.

Products

The desired product of a RAFT polymerization is typically linear polymer with an R-group at one end and a dithiocarbonate moiety at the other end. Figure depicts the major and minor products of a RAFT polymerization. All other products arise from (a) biradical termination events or (b) reactions of chemical species that originate from initiator fragments, denoted by I in the figure. (Note that categories (a) and (b) intersect).

The selectivity towards the desired product can be increased by increasing the concentration of RAFT agent relative to the quantity of free radicals delivered during the polymerization. This can be done either directly (i.e. by increasing the RAFT agent concentration) or by decreasing the rate of decomposition of or concentration of initiator.

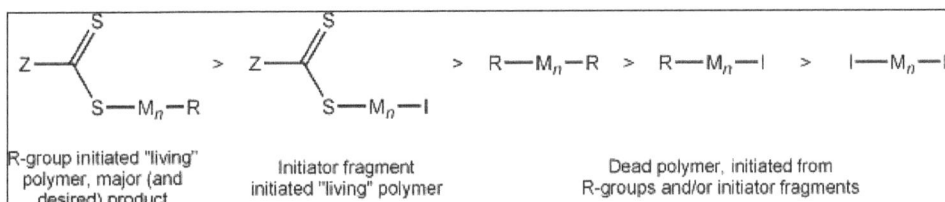

Major product of a RAFT polymerization (left) and other biproducts, arranged in order of decreasing prevalence.

RAFT Mechanism

Kinetics Overview

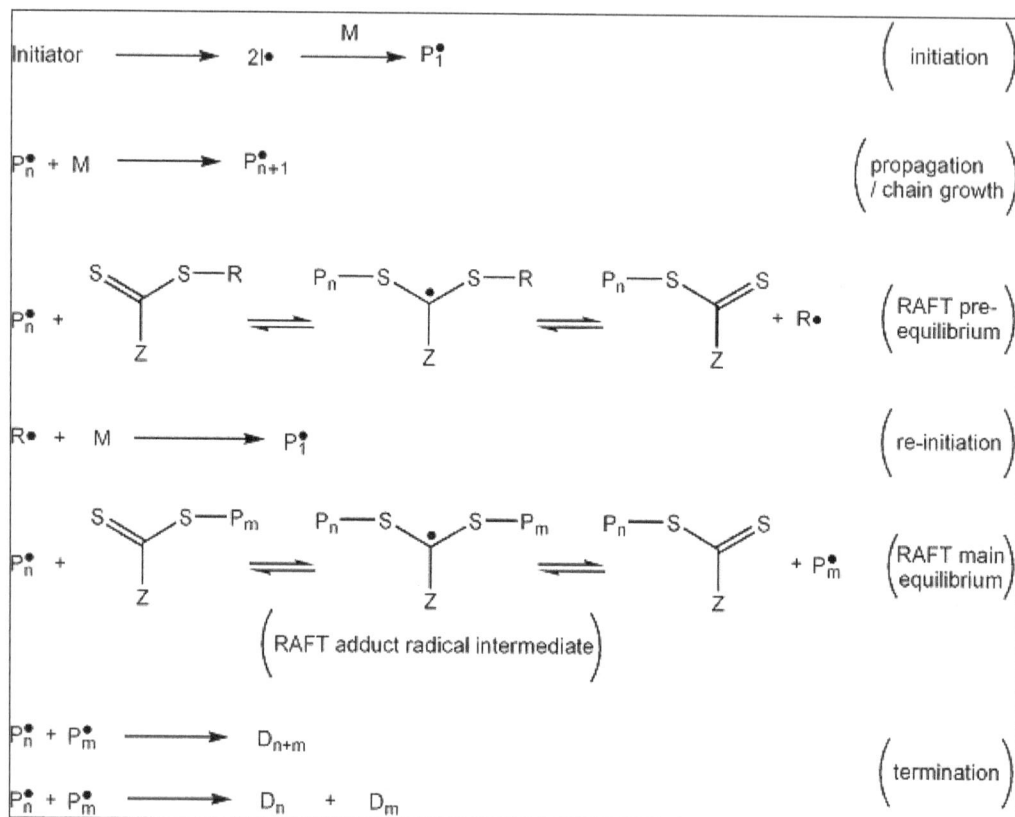

Simplified mechanism of the RAFT process.

RAFT is a type of living polymerization involving a conventional radical polymerization which is mediated by a RAFT agent. Monomers must be capable of radical polymerization. There are a number of steps in a RAFT polymerization: initiation, pre-equilibrium, re-initiation, main equilibrium, propagation and termination. The mechanism is now explained further with the help of figure.

- Initiation: The reaction is started by a free-radical source which may be a decomposing radical initiator such as AIBN. In the example in figure, the initiator decomposes to form two fragments (I•) which react with a single monomer molecule to yield a propagating (i.e. growing) polymeric radical of length 1, denoted P_1•.

- Propagation: Propagating radical chains of length n in their active (radical) form, P_n•, add to monomer, M, to form longer propagating radicals, P_{n+1}•.

- RAFT pre-equilibrium: A polymeric radical with n monomer units (P_n) reacts with the RAFT agent to form a RAFT adduct radical. This may undergo a fragmentation reaction in either direction to yield either the starting species or a radical (R•) and a polymeric RAFT agent (S=C(Z)S-P$_n$). This is a reversible step in which the intermediate RAFT adduct radical is capable of losing either the R group (R•) or the polymeric species (P$_n$•).

- Re-initiation: The leaving group radical (R•) then reacts with another monomer species, starting another active polymer chain.

- Main RAFT equilibrium: This is the most important part in the RAFT process, in which, by a process of rapid interchange, the present radicals (and hence opportunities for polymer chain growth) are "shared" among all species that have not yet undergone termination (P$_n$• and S=C(Z)S-P$_n$). Ideally the radicals are shared equally, causing chains to have equal opportunities for growth and a narrow PDI.

- Termination: Chains in their active form react via a process known as bi-radical termination to form chains that cannot react further, known as dead polymer. Ideally, the RAFT adduct radical is sufficiently hindered such that it does not undergo termination reactions.

Thermodynamics of the Main RAFT Equilibrium

The position of the main RAFT equilibrium is affected by the relative stabilities of the RAFT adduct radical (P$_n$-S-C•(Z)-S-P$_m$) and its fragmentation products, namely S=C(Z) S-P$_n$ and polymeric radical (P$_m$•). If formation of the RAFT adduct radical is sufficiently thermodynamically favorable, the concentration of active species, P$_m$•, will be reduced to the extent that a reduction in the rate of conversion of monomer into polymer is also observed, as compared to an equivalent polymerization without RAFT agent. Such a polymerization, is referred to as a rate-retarded RAFT polymerization.

The rate of a RAFT polymerization, that is, the rate of conversion of monomer into polymer, mainly depends on the rate of the Propagation reaction because the rate of initiation and termination are much higher than the rate of propagation. The rate of propagation is proportional to the concentration, [P•], of the active species P•, whereas the rate of the termination reaction, being second order, is proportional to the square [P•]^2. This means that during rate-retarded RAFT polymerizations, the rate of formation of termination products is suppressed to a greater extent than the rate of chain growth.

In RAFT polymerizations without rate-retardation, the concentration of the active species P• is close to that in an equivalent conventional polymerization in the absence of RAFT agent.

The main RAFT equilibrium and hence the rate retardation of the reaction is influenced by both temperature and chemical factors. A high temperature favors formation of the fragmentation products rather than the adduct radical P_n-S-C•(Z)-S-P_m. RAFT agents with a radical stabilising Z-group such as Phenyl group favor the adduct radical, as do propagating radicals whose monomers lack radical stabilising features, for example Vinyl acetate.

Further Mechanistic Considerations

In terms of mechanism, an ideal RAFT polymerization has several features. The pre-equilibrium and re-initiation steps are completed very early in the polymerization meaning that the major product of the reaction (the RAFT polymer chains, RAFT-P_n), all start growing at approximately the same time. The forward and reverse reactions of the main RAFT equilibrium are fast, favoring equal growth opportunities amongst the chains. The total number of radicals delivered to the system by the initiator during the course of the polymerization is low compared to the number of RAFT agent molecules, meaning that the R group initiated polymer chains from the re-initiation step form the majority of the chains in the system, rather than initiator fragment bearing chains formed in the Initiation step. This is important because initiator decomposes continuously during the polymerization, not just at the start, and polymer chains arising from initiator decomposition cannot, therefore, have a narrow length distribution. These mechanistic features lead to an average chain length that increases linearly with the conversion of monomer into polymer.

In contrast to other controlled radical polymerizations (for example ATRP), a RAFT polymerization does not achieve controlled evolution of molecular weight and low polydispersity by reducing bi-radical termination events (although in some systems, these events may indeed be reduced somewhat, as outlined above), but rather, by ensuring that most polymer chains start growing at approximately the same time and experience equal growth during polymerization.

Applications

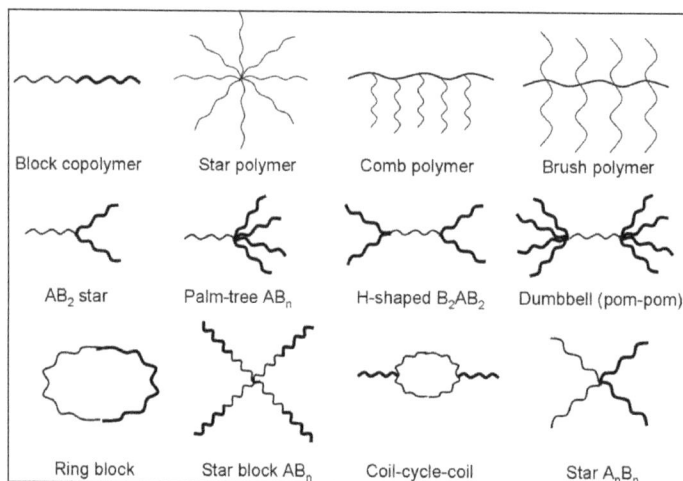

Complex architectures accessible via the RAFT process.

RAFT polymerization has been used to synthesize a wide range of polymers with controlled molecular weight and low polydispersities (between 1.05 and 1.4 for many monomers).

RAFT polymerization is known for its compatibility with a wide range of monomers as compared to other controlled radical polymerizations. Some monomers capable of polymerizing by RAFT include styrenes, acrylates, acrylamides, and many vinyl monomers. Additionally, the RAFT process allows the synthesis of polymers with specific macromolecular architectures such as block, gradient, statistical, comb, brush, star, hyperbranched, and network copolymers. These properties make RAFT useful in many types of polymer synthesis.

Block Copolymers

As with other living radical polymerization techniques, RAFT allows chain extension of a polymer of one monomer with a second type of polymer to yield a block copolymer. In such a polymerisation, there is the additional challenge that the RAFT agent for the first monomer must also be suitable for the second monomer, making block copolymerisation of monomers of highly disparate character challenging.

Multiblock copolymers have also been reported by using difunctional R groups or symmetrical trithiocarbonates with difunctional Z groups.

Star, Brush and Comb Polymers

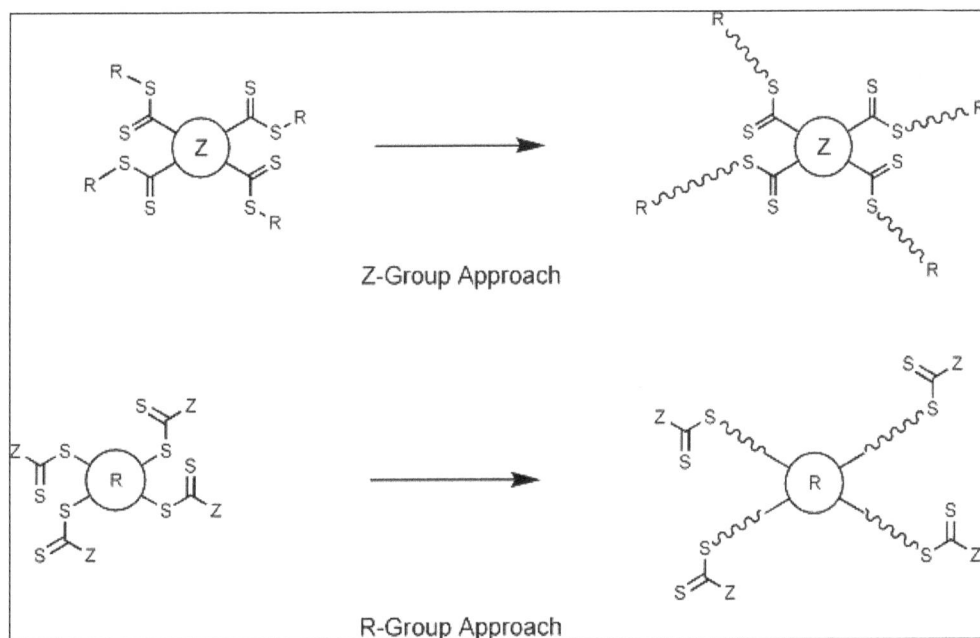

RAFT R-group approach v.s. Z-group approach.

Using a compound with multiple dithio moieties (often termed a multifunctional

RAFT agent) can result in the formation of star, brush and comb polymers. Taking star polymers as an example, RAFT differs from other forms of living radical polymerization techniques in that either the R- or Z-group may form the core of the star. While utilizing the R-group as the core results in similar structures found using ATRP or NMP, the ability to use the Z-group as the core makes RAFT unique. When the Z-group is used, the reactive polymeric arms are detached from the star's core during growth and to undergo chain transfer, must once again react at the core.

Smart Materials and Biological Applications

Due to its flexibility with respect to the choice of monomers and reaction conditions, the RAFT process competes favorably with other forms of living polymerization for the generation of bio-materials. New types of polymers are able to be constructed with unique properties, such as temperature and pH sensitivity.

Specific materials and their applications include polymer-protein and polymer-drug conjugates, mediation of enzyme activity, molecular recognition processes and polymeric micelles which can deliver a drug to a specific site in the body. RAFT has also been used to graft polymer chains onto polymeric surfaces, for example, polymeric microspheres.

RAFT Compared to other Controlled Polymerizations

Advantages

Polymerization can be performed in large range of solvents (including water), within a wide temperature range, high functional group tolerance and absence metals for polymerization. As of 2014, the range of commercially available RAFT agents covers close to all the monomer classes that can undergo radical polymerization.

Disadvantages

A particular RAFT agent is only suitable for a limited set of monomers and the synthesis of a RAFT agent typically requires a multistep synthetic procedure and subsequent purification. RAFT agents can be unstable over long time periods, are highly colored and can have a pungent odor due to gradual decomposition of the dithioester moiety to yield small sulfur compounds. The presence of sulfur and color in the resulting polymer may also be undesirable for some applications; however, this can, to an extent, be eliminated with further chemical and physical purification steps.

Catalyst Transfer Polymerization

Catalyst transfer polymerization (CTP), or Catalyst Transfer Polycondensation, is a type of living chain-growth polymerization that is used for synthesizing conjugated polymers. Benefits to using CTP over other methods are lower polydispersity and greater control over number average molecular weight in the resulting polymer sample, but very few monomers have been demonstrated to undergo CTP.

Characteristics

CTP is exclusively performed on arene monomers to give conjugated polymers. The polymers obtained from a true CTP are often low dispersity and contain consistent end groups (two on each chain, one put on the polymer chain end during initiation, and the other put on during termination). These result from the fact that CTP is a living, chain growth polymerization, and mean that the average molecular weight of the polymers from a CTP reaction can be tuned by changing the relative amounts of catalyst and monomer at the start of the reaction.

CTP is thought to be important in that control over molecular weight (imparted by CTP) means control over conjugation length in conjugated polymers. Average conjugation length of a conjugated polymer can have a large impact in applications such as solar cells and transistors, both fields that are starting to use organic electronics.

Types

CTP is closely related to many group 10 metal catalyzed cross coupling reactions and often, monomers used in CTP contain magnesium-, zinc-, boron-, and tin-based transmetallating groups, giving rise to Kumada CTP (K-CTP), Negishi CTP (N-CTP), Suzuki CTP (B-CTP), and Stille CTP (S-CTP) reactions, respectively.

Mechanism

The mechanism of CTP has been debated. The living chain-growth nature of CTP can be explained by the existence of a π-complex but can also be explained via polymer reactivity. A common feature of both pathways is their similarity to other nickel and palladium catalyzed cross-coupling reactions.

Initiation

Initiation from a nickel(II) species involves two monomers transmetalating onto the nickel center to form a complex that can easily undergo reductive elimination. The complex formed after reductive elimination is characterized by the metal, now nickel(0), bound to the π system of the monomer. This π-complex means that the catalysts is not free to dissociate and start polymer chains, forcing monomers to add to the growing chain. The catalyst migrates, via ring-walking, to the π-bond adjacent to the C-X bond at the end of the dimer, allowing oxidative addition to occur, forming the active polymer-nickel(II)-bromide catalyst.

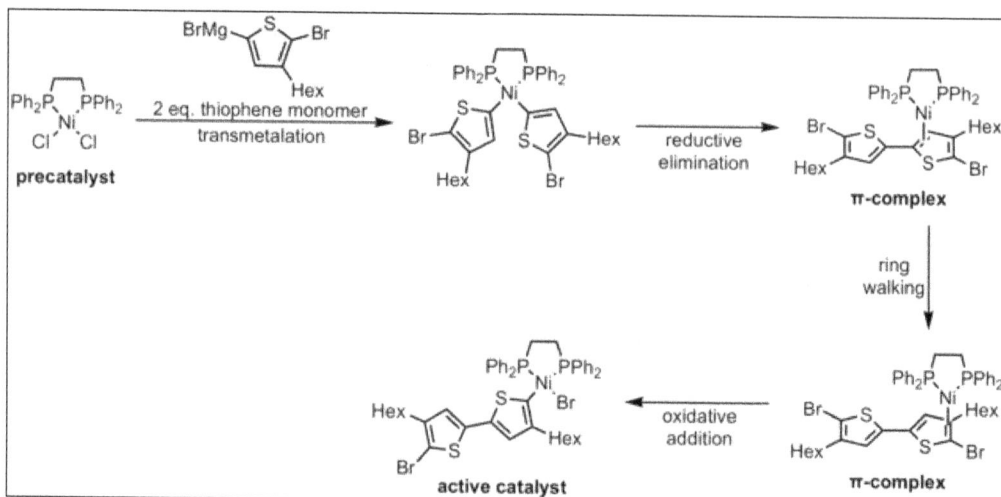

Propagation

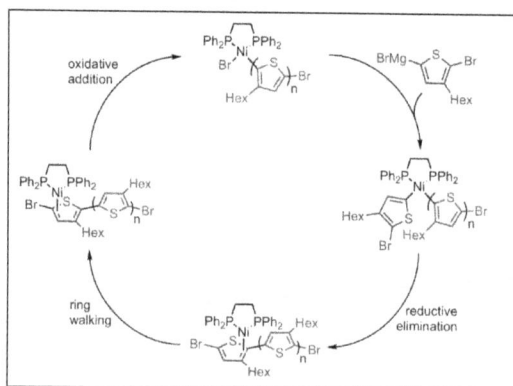

The propagation steps of CTP occurs through a similar cycle of transmetalation, reductive elimination, ring walking, and oxidative addition. The existence of a π-complex allows for the polymerization to be controlled as it ensures that the catalyst cannot dissociate from the polymer chain (and start new chains). This means that the number of polymer chains at the end of the polymerization should be equal to the number of catalysts in solution and that the average degree of polymerization in the sample at the end of polymerization should be equal to the ratio of monomers to catalysts in solution (e.g. if there are 50 equivalents of monomer to one equivalent of catalyst, most of the polymers at the end of polymerization will have 50 repeat units).

Termination

A characteristic of a true CTP is living growth character, meaning that once all monomer has been converted, addition of more monomer to the polymerization should result in monomers being added to the end of polymer chains (rather than starting new chains). This should be true of CTP, but the organometallic nature of the polymer chain end (being capped with the nickel catalyst after C-X oxidative addition), means that if the polymerization is allowed to go to high conversion, disproportionation occurs between two chain ends forming polymers with an integer multiple of the expected molecular weight. For this reason, when CTP is used to make polymers of a specific molecular weight, the polymerization is terminated early by addition of strong acid, iodine, or bromine.

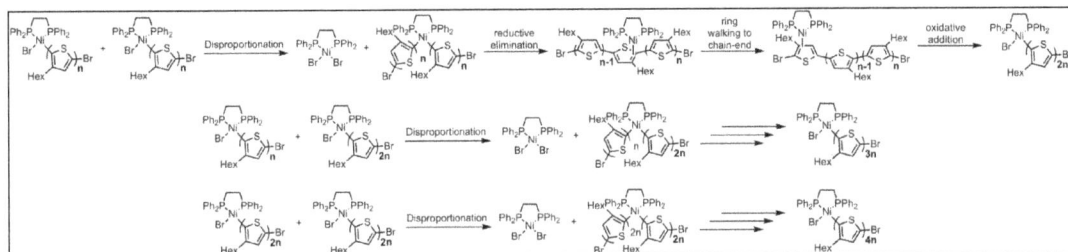

Additionally, if the π-complex is too weakly bound, termination of polymer chains can occur if a chain transfer agent (i.e. solvent, monomer, polymer, any species that can bind to the catalyst) displaces the catalyst from its polymer chain, poisoning the catalyst or starting new chains. This gives polymers with lower than expected average molecular weight.

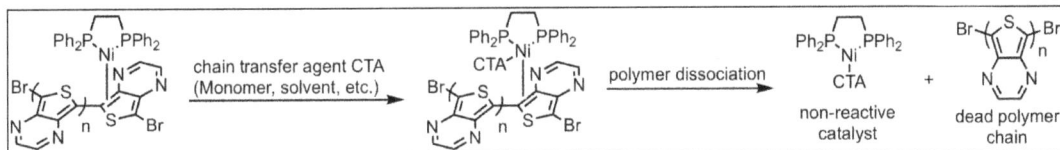

Current research into CTP focuses on finding catalysts that form strong catalyst-polymer π-complexes such that the polymerization remains living.

Analysis

Success of CTP is often evaluated using gel permeation chromatography, matrix-assisted laser desorption/ionization, nuclear magnetic resonance spectroscopy, and gas chromatography. GPC characterization allows for determination of average molecular weight and whether disproportionation has occurred (evidenced by multiple peaks in the GPC trace at different, integer multiple molecular weights of the lowest molecular weight peak). MALDI and NMR allow for identification of end groups of the polymer chain, which can help to elucidate if the polymerization is living. And GC allows for determination of conversion, and can be used in conjunction with other methods to create a molecular weight vs. monomer conversion plot.

Polymer Reactivity vs. π-complex

In theory, the chain growth nature of CTP can be described without invoking a catalyst-polymer π-complex. If we assume that no π-complex forms and instead every time a monomer was added to a polymer, the polymer becomes more reactive, we would also see chain growth since the largest polymers in the reaction would be the most reactive and would react with monomers preferentially. The difference between these two explanations of chain-growth can be elucidated by studying the presence of end groups of the polymers using mass spectrometry.

Polymers that can be Synthesized via CTP

The monomer scope of CTP is limited, but a non-exhaustive list of the polymers that can be synthesized using CTP are below:

- Polythiophene.

- Polyphenylene.

- Polyselenophene.

- Polytellurophene.

- Polythiazole.

- Polybenzothiadiazole.

- Poly(1-hexene)-block-poly(thiophene) (with the 1-hexene block polymerizeed via chain growth, and thiophene polymerized via CTP).

- Polypyrrole.

- Polyfluorene.

Cationic Polymerization

Cationic polymerization is a type of chain growth polymerization in which a cationic initiator transfers charge to a monomer which then becomes reactive. This reactive monomer goes on to react similarly with other monomers to form a polymer. The types of monomers necessary for cationic polymerization are limited to alkenes with electron-donating substituents and heterocycles. Similar to anionic polymerization reactions, cationic polymerization reactions are very sensitive to the type of solvent used. Specifically, the ability of a solvent to form free ions will dictate the reactivity of the propagating cationic chain. Cationic polymerization is used in the production of polyisobutylene (used in inner tubes) and poly(N-vinylcarbazole) (PVK).

Monomers

Monomer scope for cationic polymerization is limited to two main types: alkene and heterocyclic monomers. Cationic polymerization of both types of monomers occurs only if the overall reaction is thermally favorable. In the case of alkenes, this is due to isomerization of the monomer double bond; for heterocycles, this is due to release of monomer ring strain and, in some cases, isomerization of repeating units. Monomers for cationic polymerization are nucleophilic and form a stable cation upon polymerization.

Alkenes

Cationic polymerization of olefin monomers occurs with olefins that contain electron-donating substituents. These electron-donating groups make the olefin nucleophilic enough to attack electrophilic initiators or growing polymer chains. At the same time, these electron-donating groups attached to the monomer must be able to stabilize the resulting cationic charge for further polymerization. Some reactive olefin monomers are shown below in order of decreasing reactivity, with heteroatom groups being more reactive than alkyl or aryl groups. Note, however, that the reactivity of the carbenium ion formed is the opposite of the monomer reactivity.

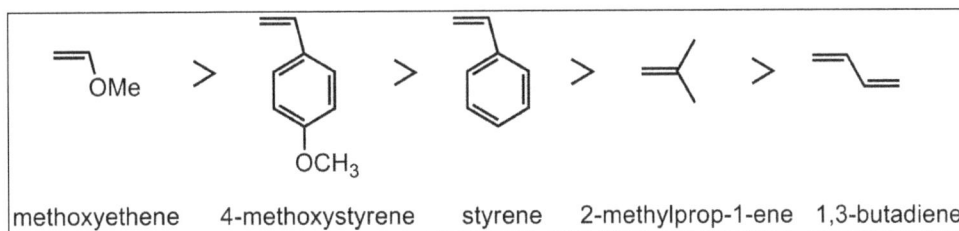

| methoxyethene | 4-methoxystyrene | styrene | 2-methylprop-1-ene | 1,3-butadiene |

Decreasing reactivity of alkene monomers.

Heterocyclic Monomers

Heterocyclic monomers that are cationically polymerized are lactones, lactams and cyclic amines. Upon addition of an initiator, cyclic monomers go on to form linear polymers. The reactivity of heterocyclic monomers depends on their ring strain. Monomers with large ring strain, such as oxirane, are more reactive than 1,3-dioxepane which has considerably less ring strain. Rings that are six-membered and larger are less likely to polymerize due to lower ring strain.

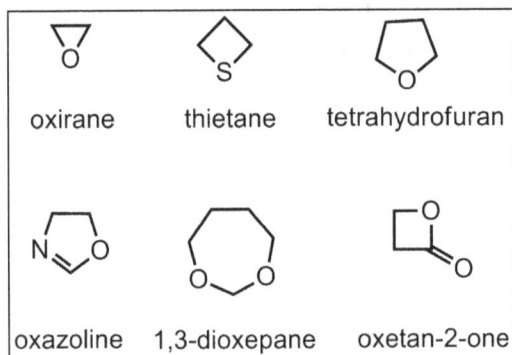

| oxirane | thietane | tetrahydrofuran |
| oxazoline | 1,3-dioxepane | oxetan-2-one |

Examples of heterocyclic monomers.

Synthesis

Initiation

Initiation is the first step in cationic polymerization. During initiation, a carbenium ion is generated from which the polymer chain is made. The counterion should be non-nucleophilic, otherwise the reaction is terminated instantaneously. There are a variety of initiators available for cationic polymerization, and some of them require a coinitiator to generate the needed cationic species.

Classical Protic Acids

Strong protic acids can be used to form a cationic initiating species. High concentrations of the acid are needed in order to produce sufficient quantities of the cationic species. The counterion (A^-) produced must be weakly nucleophilic so as to prevent early termination due to combination with the protonated alkene. Common acids used are phosphoric, sulfuric, fluoro-, and triflic acids. Only low molecular weight polymers are formed with these initiators.

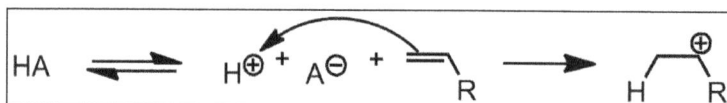

Initiation by protic acids

Lewis Acids/Friedel-crafts Catalysts

Lewis acids are the most common compounds used for initiation of cationic polymerization. The more popular Lewis acids are $SnCl_4$, $AlCl_3$, BF_3, and $TiCl_4$. Although these Lewis acids alone are able to induce polymerization, the reaction occurs much faster with a suitable cation source. The cation source can be water, alcohols, or even a carbocation donor such as an ester or an anhydride. In these systems the Lewis acid is referred to as a coinitiator while the cation source is the initiator. Upon reaction of the initiator with the coinitiator, an intermediate complex is formed which then goes on to react with the monomer unit. The counterion produced by the initiator-coinitiator complex is less nucleophilic than that of the Brønsted acid A^- counterion. Halogens, such as chlorine and bromine, can also initiate cationic polymerization upon addition of the more active Lewis acids.

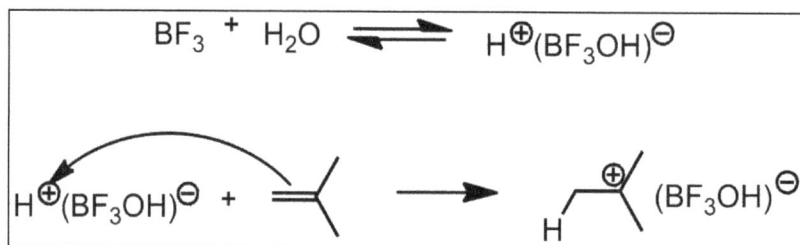

Initiation with boron trifluoride (coinitiator) and water (initiator).

Carbenium Ion Salts

Stable carbenium ions are used to initiate chain growth of only the most reactive alkenes and are known to give well defined structures. These initiators are most often used in kinetic studies due to the ease of measuring the disappearance of the carbenium ion absorbance. Common carbenium ions are trityl and tropylium cations.

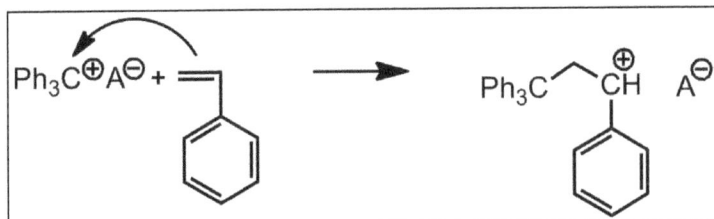

Initiation with trityl carbenium ion.

Ionizing Radiation

Ionizing radiation can form a radical-cation pair that can then react with a monomer to start cationic polymerization. Control of the radical-cation pairs are difficult and often depend on the monomer and reaction conditions. Formation of radical and anionic species are often observed.

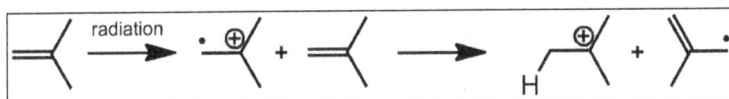

Initiation using ionizing radiation.

Propagation

Propagation proceeds by addition of monomer to the active species, i.e. the carbenium ion. The monomer is added to the growing chain in a head-to-tail fashion; in the process, the cationic end group is regenerated to allow for the next round of monomer addition.

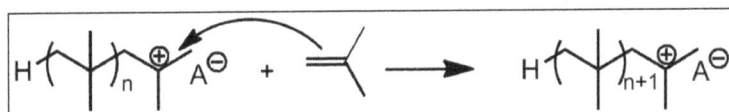

General propagation pathway.

Effect of Temperature

The temperature of the reaction has an effect on the rate of propagation. The overall activation energy for the polymerization (E) is based upon the activation energies for the initiation (E_i), propagation (E_p), and termination (E_t) steps:

$$E = E_i + E_p - E_t$$

Generally, E_t is larger than the sum of E_i and E_p, meaning the overall activation energy is negative. When this is the case, a decrease in temperature leads to an increase in the rate of propagation. The converse is true when the overall activation energy is positive.

Chain length is also affected by temperature. Low reaction temperatures, in the range of 170–190 K, are preferred for producing longer chains. This comes as a result of the activation energy for termination and other side reactions being larger than the activation energy for propagation. As the temperature is raised, the energy barrier for the termination reaction is overcome, causing shorter chains to be produced during the polymerization process.

Effect of Solvent and Counterion

The solvent and the counterion (the gegen ion) have a significant effect on the rate of propagation. The counterion and the carbenium ion can have different associations according to intimate ion pair theory; ranging from a covalent bond, tight ion pair (unseparated), solvent-separated ion pair (partially separated), and free ions (completely dissociated). Range of associations between the carbenium ion (R^+) and gegen ion (X^-):

$\sim\sim\sim RX$ covalent $\sim\sim\sim R^+X^-$ tight ion pair $\sim\sim\sim R^{+}/X^-$ solvent-separated ion pair $\sim\sim\sim R^+ + X^-$ free ions

The association is strongest as a covalent bond and weakest when the pair exists as free ions. In cationic polymerization, the ions tend to be in equilibrium between an ion pair (either tight or solvent-separated) and free ions. The more polar the solvent used in the reaction, the better the solvation and separation of the ions. Since free ions are more reactive than ion pairs, the rate of propagation is faster in more polar solvents.

The size of the counterion is also a factor. A smaller counterion, with a higher charge density, will have stronger electrostatic interactions with the carbenium ion than will a larger counterion which has a lower charge density. Further, a smaller counterion is more easily solvated by a polar solvent than a counterion with low charge density. The result is increased propagation rate with increased solvating capability of the solvent.

Termination

Termination generally occurs by unimolecular rearrangement with the counterion. In this process, an anionic fragment of the counterion combines with the propagating chain end. This not only inactivates the growing chain, but it also terminates the kinetic chain by reducing the concentration of the initiator-coinitiator complex.

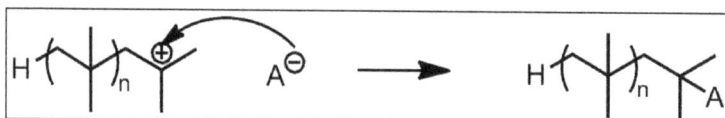

Termination by combination with an anionic fragment from the counterion.

Chain Transfer

Chain transfer can take place in two ways. One method of chain transfer is hydrogen abstraction from the active chain end to the counterion. In this process, the growing chain is terminated, but the initiator-coinitiator complex is regenerated to initiate more chains.

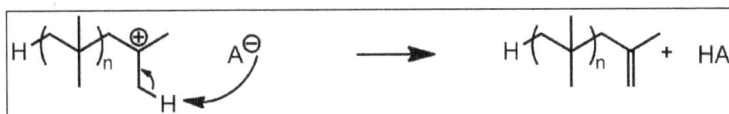

Chain transfer by hydrogen abstraction to the counterion.

The second method involves hydrogen abstraction from the active chain end to the monomer. This terminates the growing chain and also forms a new active carbenium ion-counterion complex which can continue to propagate, thus keeping the kinetic chain intact.

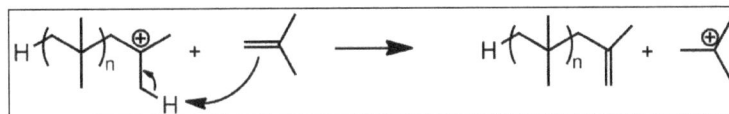

Chain transfer by hydrogen abstraction to the monomer.

Cationic Ring-opening Polymerization

Cationic ring-opening polymerization follows the same mechanistic steps of initiation, propagation, and termination. However, in this polymerization reaction, the monomer units are cyclic in comparison to the resulting polymer chains which are linear. The linear polymers produced can have low ceiling temperatures, hence end-capping of the polymer chains is often necessary to prevent depolymerization.

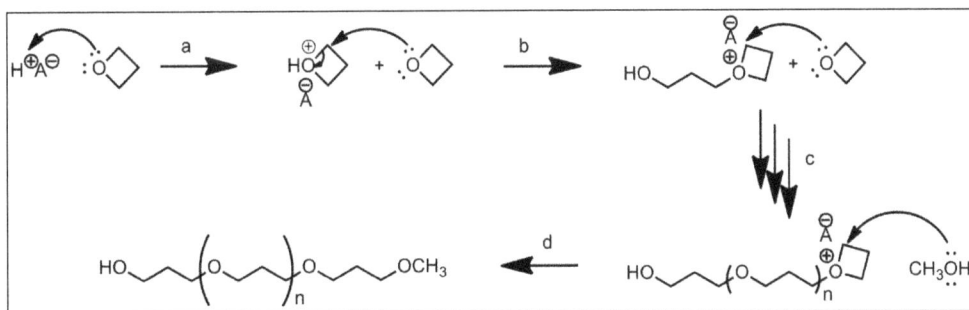

Cationic ring-opening polymerization of oxetane involving
(a and b) initiation, (c) propagation, and (d) termination with methanol.

Kinetics

The rate of propagation and the degree of polymerization can be determined from an analysis of the kinetics of the polymerization. The reaction equations for initiation, propagation, termination, and chain transfer can be written in a general form:

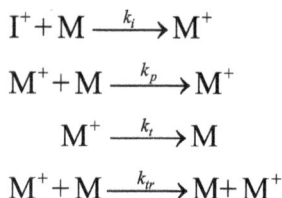

$$I^+ + M \xrightarrow{k_i} M^+$$

$$M^+ + M \xrightarrow{k_p} M^+$$

$$M^+ \xrightarrow{k_t} M$$

$$M^+ + M \xrightarrow{k_{tr}} M + M^+$$

In which I^+ is the initiator, M is the monomer, M^+ is the propagating center, and k_i, k_p, , and k_{tr} are the rate constants for initiation, propagation, termination, and chain transfer, respectively. For simplicity, counterions are not shown in the above reaction equations and only chain transfer to monomer is considered. The resulting rate equations are as follows, where brackets denote concentrations:

$$\text{rate(initiation)} = k_i[I^+][M]$$

$$\text{rate(propagation)} = k_p[M^+][M]$$

$$\text{rate(termination)} = k_t[M^+]$$

$$\text{rate(chain transfer)} = k_{tr}[M^+][M]$$

Assuming steady-state conditions, i.e. the rate of initiation = rate of termination:

$$[M^+] = \frac{k_i[I^+][=M]}{k_t}$$

This equation for $[M^+]$ can then be used in the equation for the rate of propagation:

$$\text{rate(propagation)} = \frac{k_p k_i[M]^2[I^+]}{k_t}$$

From this equation, it is seen that propagation rate increases with increasing monomer and initiator concentration.

The degree of polymerization, X_n, can be determined from the rates of propagation and termination:

$$X_n = \frac{\text{rate(propagation)}}{\text{rate(termination)}} = \frac{k_p[M]}{k_t}$$

If chain transfer rather than termination is dominant, the equation for X_n becomes:

$$X_n = \frac{\text{rate(propagation)}}{\text{rate(chain transfer)}} = \frac{k_p}{k_{tr}}$$

Living Polymerization

In 1984, Higashimura and Sawamoto reported the first living cationic polymerization for alkyl vinyl ethers. This type of polymerization has allowed for the control of well-defined polymers. A key characteristic of living cationic polymerization is that termination is essentially eliminated, thus the cationic chain growth continues until all monomer is consumed.

Commercial Applications

The largest commercial application of cationic polymerization is in the production of polyisobutylene (PIB) products which include polybutene and butyl rubber. These polymers have a variety of applications from adhesives and sealants to protective gloves and pharmaceutical stoppers. The reaction conditions for the synthesis of each type of isobutylene product vary depending on the desired molecular weight and what type(s) of monomer(s) is used. The conditions most commonly used to form low molecular weight ($5-10 \times 10^4$ Da) polyisobutylene are initiation with $AlCl_3$, BF_3, or $TiCl_4$ at a temperature range of -40 to $10\ °C$. These low molecular weight polyisobutylene polymers are used for caulking and as sealants. High molecular weight PIBs are synthesized at much lower temperatures of -100 to $-90\ °C$ and in a polar medium of methylene chloride. These polymers are used to make uncrosslinked rubber products and are additives for certain thermoplasts. Another characteristic of high molecular weight PIB is its low toxicity which allows it to be used as a base for chewing gum. The main chemical companies that produce polyisobutylene are Esso, ExxonMobil, and BASF.

Butyl rubber gloves.

Butyl rubber, in contrast to PIB, is a copolymer in which the monomers isobutylene (~98%) and isoprene (2%) are polymerized in a process similar to high molecular weight PIBs. Butyl rubber polymerization is carried out as a continuous process with $AlCl_3$ as the initiator. Its low gas permeability and good resistance to chemicals and aging make it useful for a variety of applications such as protective gloves, electrical cable insulation, and even basketballs. Large scale production of butyl rubber started during World War II, and roughly 1 billion pounds/year are produced in the U.S. today.

Polybutene is another copolymer, containing roughly 80% isobutylene and 20% other butenes (usually 1-butene). The production of these low molecular weight polymers (300–2500 Da) is done within a large range of temperatures (−45 to 80 °C) with $AlCl_3$ or BF_3. Depending on the molecular weight of these polymers, they can be used as adhesives, sealants, plasticizers, additives for transmission fluids, and a variety of other applications. These materials are low-cost and are made by a variety of different companies including BP Chemicals, Esso, and BASF.

Other polymers formed by cationic polymerization are homopolymers and copolymers of polyterpenes, such as pinenes (plant-derived products), that are used as tackyfiers. In the field of heterocycles, 1,3,5-trioxane is copolymerized with small amounts of ethylene oxide to form the highly crystalline polyoxymethylene plastic. Also, the homopolymerization of alkyl vinyl ethers is achieved only by cationic polymerization.

Cobalt-mediated Radical Polymerization

Cobalt based catalysts, when used in radical polymerization, have several main advantages especially in slowing down the reaction rate, allowing for the synthesis of polymers with peculiar properties. As starting the reaction does need a real radical initiator, the cobalt species is not the only used catalyst, it is a mediator. For this reason this type of reaction is referred to as cobalt mediated radical polymerization.

About half of all commercial polymers are produced by radical polymerization. Radical polymerization (RP) reactions have several advantageous properties:

- A wide variety of monomers can be polymerized.
- RP reactions are tolerant to various functional groups.
- RP reactions allow a large temperature range of operation (−100 to >200 °C).
- RP reactions are generally compatibility with several reactions conditions (bulk, solution, (mini)emulsion and suspension).
- RP reactions allow a relatively simple reactor set-up, and are hence costs effective.

However, conventional (free) RP reactions suffer from a lack of control over the polymer molecular-weights and weight distributions. A relatively narrow molecular weight-distribution (M_w/M_n) is usually desirable, as broad distribution negatively influence the polymer properties of (generally atactic) polymers produced by RP. Common RP also does not allow the formation of block copolymers. This is where controlled (or living) radical polymerization comes into play. Several CRP reactions have been developed over the past years, some of which capable of producing well-defined polymers with narrow molecular weight distributions.

Cobalt mediated radical polymerisation (CMRP) is one of these methods, which offers some specific advantages. Most notably, CMRP allows RP of a broad substrate scope (among others acrylates, acrylic acid, vinyl esters, vinyl acetate, acrylonitrile, vinylpyrrolidone) under various reaction conditions, and (for some catalysts) gives access to very fast CRP reactions with rates approaching those of conventional uncontrolled free radical polymerization reactions.

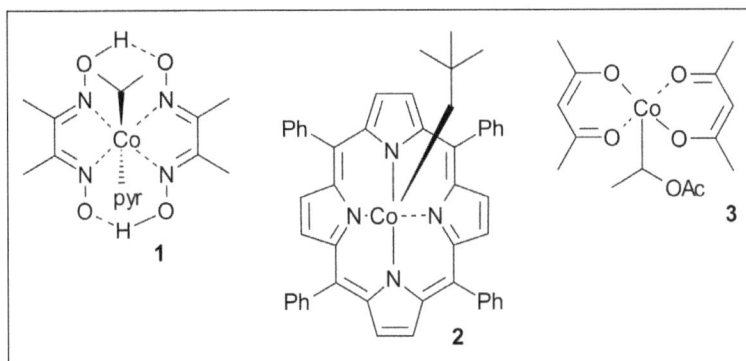

Most commonly applied cobalt compounds are cobaloximes, cobalt porphyrins and Co(acac)2 derivatives, used in combination with various radical initiators (such as AIBN or V70).

Cobalt can control radical polymerization (RP) reactions by essentially three mechanisms:

1. Catalytic chain transfer (CCT).

2. Reversibile termination (RT), leading to the persistent radical effect (PRE).

3. Degenerative transfer (DT).

Control via Reversible Termination (Persistent Radical Effect)

In many cases, CMRP exploits the weak cobalt(III)-carbon bond to control the radical polymerization reaction. The Co-C bond containing radical initiator easily breaks up (by heat or by light) in a carbon free radical and a cobalt(II) radical species. The carbon radical starts the growth of a polymer chain from the $CH_2 = CHX$ monomer as in a free radical polymerization reaction. Cobalt is unusual in that it can reversibly reform a covalent bond with the carbon radical terminus of the growing chain. This reduces the concentration of radicals to a minimum and thereby minimizes undesirable termination reactions by recombination of two carbon radicals. The cobalt trapping reagent is called a persistent radical and the cobalt-capped polymer chain is said to be dormant.

This mechanism is called reversible termination and is said to operate via the "persistent radical effect". When the monomer lacks protons that can be easily abstracted by the cobalt radical, (catalytic) chain transfer is also limited and the RP reaction becomes close to 'living'.

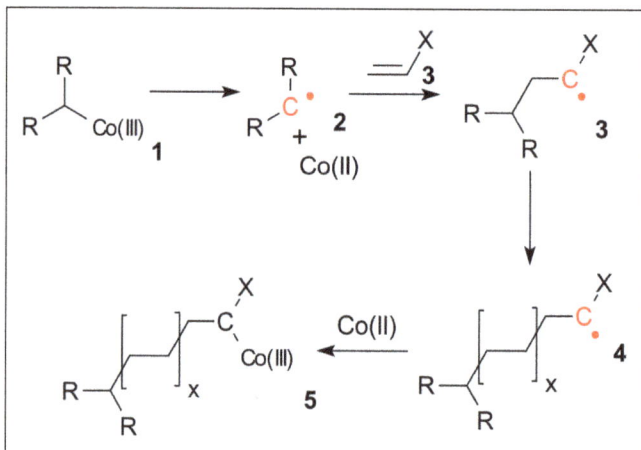

Control via Catalytic Chain Transfer

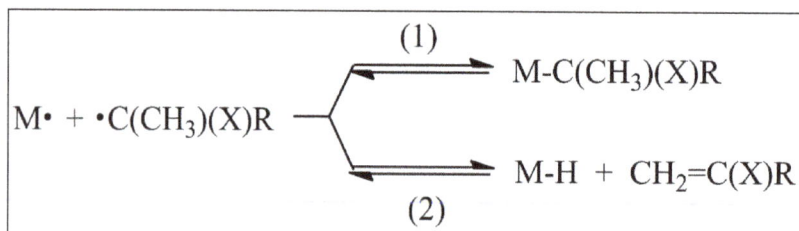

$$M\bullet + \bullet C(CH_3)(X)R \underset{(2)}{\overset{(1)}{\rightleftharpoons}} \begin{array}{l} M\text{-}C(CH_3)(X)R \\ M\text{-}H + CH_2{=}C(X)R \end{array}$$

Catalytic chain transfer is a way to make shorter polymer chains in a radical polymerization process. The method involves adding a catalytic chain transfer agent to the reaction mixture of the monomer and the radical initiator. Catalytic chain transfer proceeds via hydrogen atom transfer from the organic growing polymeryl radical to cobalt(II), producing a polymer vinyl-end group and a cobalt-hydride species (equilibrium 2). The Co-H species then reacts with the monomer to start a new Co(III)-alkyl species, which re-initiates a new growing polymeric radical (reversible termination, equilibrium 1). The main products of catalytic chain transfer polymerization are therefore vinyl terminated polymer chains which are shorter than in conventional (free) radical polymerization.

Control via Degenerative Transfer

One of the disadvantages of controlled radical polymerization reactions is that they tend to become rather slow. Controlled polymerization conditions are usually achieved by extending the life-time of the growing polymer chain radical, by keeping it in a dormant state for most of the time (known as the Persistent Radical Effect). Thereby the control agent substantially slows-down the over-all radical polymerisation reaction. However, some CMRP reactions proceed via a different mechanism, called degenerative transfer (DT), which allows controlled radical polymerization reactions to proceed at roughly the same rate as any uncontrolled free radical polymerisation.

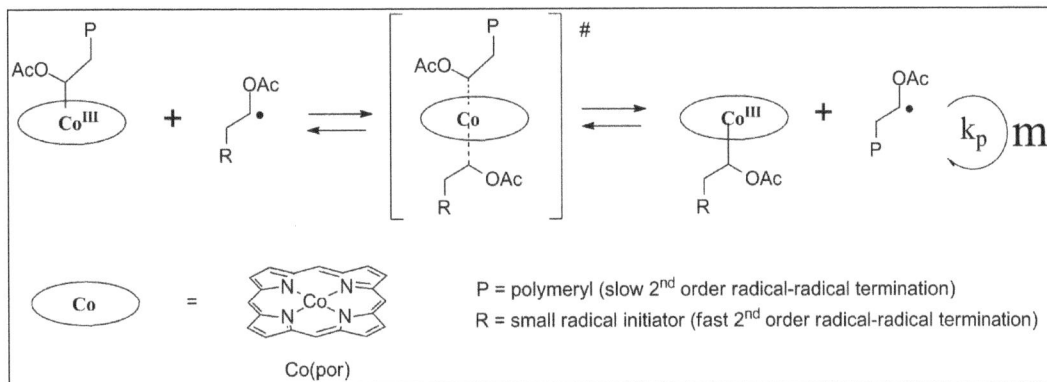

The degenerative transfer mechanism is based on very fast exchange equilibria between small free radicals (being continuously injected into the solution) and dormant polymeryl radicals (protected as closed-shell cobalt species). Systems based on degenerative transfer do not proceed via the persistent radical effect (PRE). Instead an active propagating radical interchanges its role with a latent radical in a dormant complex. The activation of one polymer chain means the deactivation of another polymer chain. If the exchange process is much faster than the polymerisation rate (k_p), effectively all polymer chains grow at the same rate. Because the large polymer chains diffuse much slower than the small organic radicals, and thereby terminate much slower via 2nd order radical-radical coupling or disproportionation, long chains effectively build-up at cobalt while the small radicals keep terminating. This leads to a desirable narrow molecular weight distribution of the polymer at high polymerization rates. DT-CMRP is an associative process, which for $Co^{III}(por)(alkyl)$ species implies formation of a 6-coordinate intermediate or transition state. Such $Co(por)(alkyl)_2$ species formally have a $Co(+IV)$ oxidation state, but in reality their (electronic) structure is best described as a weak radical adduct of a $Co^{III}(por)(alkyl)$ species. A striking feature of DT-CMRP is the fact that even upon using a large excess of the radical initiator compared to the transfer agent, the radical polymerization reactions still remains controlled. A satisfactory explanation for this phenomenon seems to be lacking at the moment.

Dispersion Polymerization

In polymer science, dispersion polymerization is a heterogeneous polymerization process carried out in the presence of a polymeric stabilizer in the reaction medium. Dispersion polymerization is a type of precipitation polymerization, meaning the solvent selected as the reaction medium is a good solvent for the monomer and the initiator, but is a non-solvent for the polymer. As the polymerization reaction proceeds, particles of polymer form, creating a non-homogeneous solution. In dispersion polymerization these particles are the locus of polymerization, with monomer being added to the particle throughout the reaction. In this sense, the mechanism for polymer formation and growth has features similar to that of emulsion polymerization. With typical precipitation polymerization, the continuous phase (the solvent solution) is the main locus of polymerization, which is the main difference between precipitation and dispersion.

Polymerization Mechanism

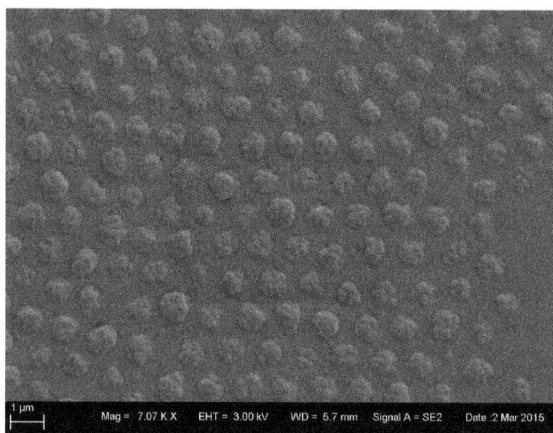

SEM-Picture of PMMA-particles fabricated by dispersion polymerization after drying/removal of the organic liquid phase (cyclohexane).

At the onset of polymerization, polymers remain in solution until they reach a critical molecular weight (MW), at which point they precipitate. These initial polymer particles are unstable and coagulate with other particles until stabilized particles form. After this point in the polymerization, growth only occurs by addition of monomer to the stabilized particles. As the polymer particles grow, stabilizer (or dispersant) molecules attach covalently to the surface. These stabilizer molecules are generally graft or block copolymers, and can be preformed or can form *in situ* during the reaction. Typically, one side of the stabilizer copolymer has an affinity for the solvent while the other side has an affinity for the polymer particle being formed. These molecules play a crucial role in dispersion polymerization by forming a "hairy layer" around the particles that prevents particle coagulation. This controls size and colloidal stability of the particles in the reaction system. The driving force for the particle separation is steric hindrance between the outward-facing tails of the stabilizer layers.

Dispersion polymerization can produce nearly monodisperse polymer particles of 0.1–15 micrometers (μm). This is important because it fills the gap between particle size generated by conventional emulsion polymerization (0.006–0.7 μm) in batch process and that of suspension polymerization (50–1000 μm).

Applications

Particles produced by dispersion polymerization are used in a wide variety of applications. Toners, instrument calibration standards, chromatography column packing materials, liquid crystal display spacers, and biomedical and biochemical analysis all use these micron-size monodisperse particles, particles which were hard to come by before the development of dispersion polymerization methods. The dispersions are also used as surface coatings. Unlike solution coatings, dispersion coatings have viscosities that are independent of polymer MW. The viscosities of dispersions are advantageously lower than those of solutions with practical polymer levels. This allows for easier application of the coating.

One dispersion polymerization system being studied is the use of supercritical liquid carbon dioxide ($scCO_2$) as a solvent. Because of its unique solvent properties, supercritical CO_2 is an ideal medium for dispersion polymerization for many soluble-monomer with insoluble-polymer systems. For example, polymers can be separated by releasing the high pressure under which the $scCO_2$ is held. This process is more efficient than typical drying processes. Also, the principles of dispersion polymerization with $scCO_2$ follows principles of green chemistry: low solvent toxicity, low waste, efficient atom economy, and avoidance of purification steps.

Step-growth Polymerization

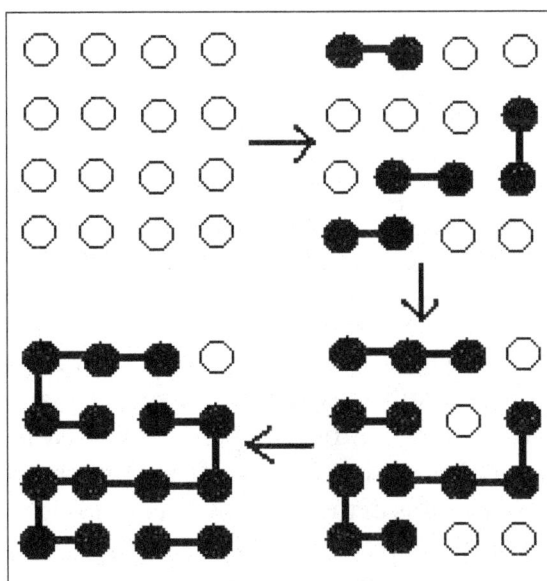

A generic representation of a step-growth polymerization. (Single white dots represent monomers and black chains represent oligomers and polymers).

Step-growth polymerization refers to a type of polymerization mechanism in which bi-functional or multifunctional monomers react to form first dimers, then trimers, longer oligomers and eventually long chain polymers. Many naturally occurring and some synthetic polymers are produced by step-growth polymerization, e.g. polyesters, polyamides, polyurethanes, etc. Due to the nature of the polymerization mechanism, a high extent of reaction is required to achieve high molecular weight. The easiest way to visualize the mechanism of a step-growth polymerization is a group of people reaching out to hold their hands to form a human chain—each person has two hands (= reactive sites). There also is the possibility to have more than two reactive sites on a monomer: In this case branched polymers production take place.

IUPAC deprecates the term step-growth polymerization and recommends use of the terms polyaddition, when the propagation steps are addition reactions and no molecules are evolved during these steps, and polycondensation when the propagation steps are condensation reactions and molecules are evolved during these steps.

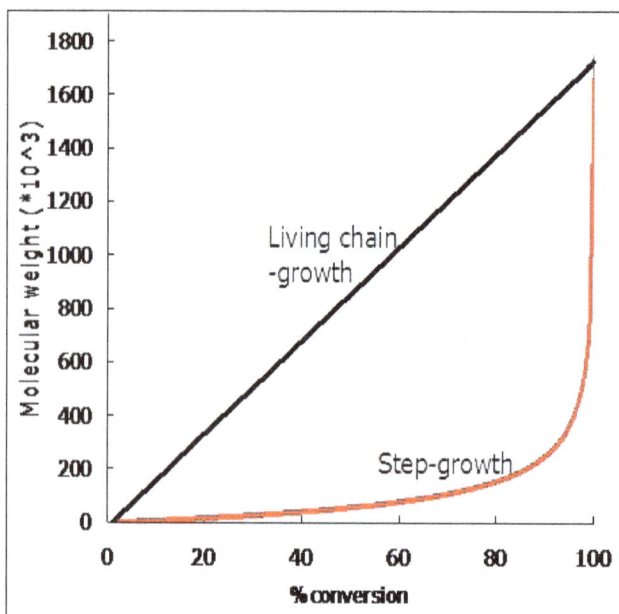

Comparison of Molecular weight vs conversion plot between step-growth and living chain-growth polymerization.

Most natural polymers being employed at early stage of human society are of condensation type. The synthesis of first truly synthetic polymeric material, Bakelite, was announced by Leo Baekeland in 1907, through a typical step-growth polymerization fashion of phenol and formaldehyde. The pioneer of synthetic polymer science, Wallace Carothers, developed a new means of making polyesters through step-growth polymerization in 1930s as a research group leader at DuPont. It was the first reaction designed and carried out with the specific purpose of creating high molecular weight polymer molecules, as well as the first polymerization reaction whose results had been predicted beforehand by scientific theory. Carothers developed a series of mathematic equations to describe the behavior of step-growth polymerization systems which are still known as the Carothers equations today. Collaborating with Paul Flory, a physical chemist, they developed theories that describe more mathematical aspects of step-growth polymerization including kinetics, stoichiometry, and molecular weight distribution etc. Carothers is also well known for his invention of Nylon.

Condensation Polymerization

"Step growth polymerization" and condensation polymerization are two different concepts, not always identical. In fact polyurethane polymerizes with addition polymerization (because its polymerization produces no small molecules), but its reaction mechanism corresponds to a step-growth polymerization.

The distinction between "addition polymerization" and "condensation polymerization" was introduced by Wallace Hume Carothers in 1929, and refers to the type of products, respectively:

- A polymer only (addition).

- A polymer and a molecule with a low molecular weight (condensation).

The distinction between "step-growth polymerization" and "chain-growth polymerization" was introduced by Paul Flory in 1953, and refers to the reaction mechanisms, respectively:

- By functional groups (step-growth polymerization).

- By free-radical or ion (chain-growth polymerization).

Differences from Chain-growth Polymerization

This technique is usually compared with chain-growth polymerization to show its characteristics.

Step-growth polymerization	Chain-growth polymerization
Growth throughout matrix.	Growth by addition of monomer only at one end or both ends of chain.
Rapid loss of monomer early in the reaction.	Some monomer remains even at long reaction times.
Similar steps repeated throughout reaction process.	Different steps operate at different stages of mechanism (i.e. Initiation, propagation, termination, and chain transfer).
Average molecular weight increases slowly at low conversion and high extents of reaction are required to obtain high chain length.	Molar mass of backbone chain increases rapidly at early stage and remains approximately the same throughout the polymerization.
Ends remain active (no termination).	Chains not active after termination.
No initiator necessary.	Initiator required.

Classes of Step-growth Polymers

Polymer class/ linear, branched A/B type	Monomers
Polyester linear $A_2 + B_2$	terephthalic acid ethylene glycol
Polyester branched $A_3 + A_2 + B_2$	benzene-1,3,5-tricarboxylic acid terephthalic acid ethylene glycol
Polyurethane branched $A_2 + B_3$ $+ B_6$ $+ B_9$	Methylene diphenyl diisocyanate 2,2',2"-((1,3,5-triazine-2,4,6-triyl) tris(azanediyl))tris(ethan-1-ol) 2,2',2",2'",2"",2"'''-((1,3,5-triazine-2,4,6-triyl) tris(azanetriyl))hexakis(ethan-1-ol) ((1,3,5-triazine-2,4,6-triyl) tris(azanediyl))tri(methanetriol)
Polyurea branched AB_2	5-isocyanatobenzene-1,3-diamine
Polyamide branched $A_2 + B_3 + B_1$	(Z)-4,4'-(diazene-1,2-diyl) bis(3-methylbenzoyl chloride) 5'-(4-aminophenyl) [1,1':3',1"-terphenyl]- 4,4"-diamine N1-(4-aminophenyl)-N4- (4-(phenylamino)phenyl) benzene-1,4-diamine

Examples of monomer systems that undergo step-growth polymerisation.
The reactive functional groups are highlighted.

Classes of step-growth polymers are:

- Polyester has high glass transition temperature T_g and high melting point T_m, good mechanical properties to about 175 °C, good resistance to solvent and chemicals. It can exist as fibers and films. The former is used in garments, felts, tire cords, etc. The latter appears in magnetic recording tape and high grade films.

- Polyamide (nylon) has good balance of properties: high strength, good elasticity and abrasion resistance, good toughness, favorable solvent resistance. The applications of polyamide include: rope, belting, fiber cloths, thread, substitute for metal in bearings, jackets on electrical wire.

- Polyurethane can exist as elastomers with good abrasion resistance, hardness, good resistance to grease and good elasticity, as fibers with excellent rebound, as coatings with good resistance to solvent attack and abrasion and as foams with good strength, good rebound and high impact strength.

- Polyurea shows high T_g, fair resistance to greases, oils, and solvents. It can be used in truck bed liners, bridge coating, caulk and decorative designs.

- Polysiloxane are available in a wide range of physical states—from liquids to greases, waxes, resins, and rubbers. Uses of this material are as antifoam and release agents, gaskets, seals, cable and wire insulation, hot liquids and gas conduits, etc.

- Polycarbonates are transparent, self-extinguishing materials. They possess properties like crystalline thermoplasticity, high impact strength, good thermal and oxidative stability. They can be used in machinery, auto-industry, and medical applications. For example, the cockpit canopy of F-22 Raptor is made of high optical quality polycarbonate.

- Polysulfides have outstanding oil and solvent resistance, good gas impermeability, good resistance to aging and ozone. However, it smells bad, and it shows low tensile strength as well as poor heat resistance. It can be used in gasoline hoses, gaskets and places that require solvent resistance and gas resistance.

- Polyether shows good thermoplastic behavior, water solubility, generally good mechanical properties, moderate strength and stiffness. It is applied in sizing for cotton and synthetic fibers, stabilizers for adhesives, binders, and film formers in pharmaceuticals.

- Phenol formaldehyde resin (Bakelite) have good heat resistance, dimensional stability as well as good resistance to most solvents. It also shows good dielectric properties. This material is typically used in molding applications, electrical, radio, televisions and automotive parts where their good dielectric properties are of use. Some other uses include: impregnating paper, varnishes, decorative laminates for wall coverings.

- Poly-Triazole polymers are produced from monomers which bear both an alkyne and azide functional group. The monomer units are linked to each other by the a 1,2,3-triazole group; which is produced by the 1,3-Dipolar cycloaddition, also called the Azide-alkyne Huisgen cycloaddition. These polymers can take on the form of a strong resin, or a gel. With oligopeptide monomers containing a terminal alkyne and terminal azide the resulting clicked peptide polymer will be biodegradable due to action of endopeptidases on the oligopeptide unit.

Branched Polymers

A monomer with functionality of 3 or more will introduce branching in a polymer and will ultimately form a cross-linked macrostructure or network even at low fractional conversion. The point at which a tree-like topology transits to a network is known as the gel point because it is signalled by an abrupt change in viscosity. One of the earliest so-called thermosets is known as bakelite. It is not always water that is released in step-growth polymerization: in acyclic diene metathesis or ADMET dienes polymerize with loss of ethylene.

Kinetics

The kinetics and rates of step-growth polymerization can be described using a polyesterification mechanism. The simple esterification is an acid-catalyzed process in which protonation of the acid is followed by interaction with the alcohol to produce an ester and water. However, there are a few assumptions needed with this kinetic model. The first assumption is water (or any other condensation product) is efficiently removed. Secondly, the functional group reactivities are independent of chain length. Finally, it is assumed that each step only involves one alcohol and one acid:

$$\frac{1}{1-p^{n-1}} = 1 + (n-1)kt[COOH]^{n-1}$$

This is a general rate law degree of polymerization for polyesterification where n= reaction order.

Self-catalyzed Polyesterification

If no acid catalyst is added, the reaction will still proceed because the acid can act as its own catalyst. The rate of condensation at any time t can then be derived from the rate of disappearance of -COOH groups:

$$rate = \frac{-d[COOH]}{dt} = k[COOH]^2[OH]$$

The second-order [\ce{COOH}] term arises from its use as a catalyst, and k is the rate constant. For a system with equivalent quantities of acid and glycol, the functional group concentration can be written simply as:

$$rate = \frac{-d[COOH]}{dt} = k[COOH]^3$$

After integration and substitution from Carothers equation, the final form is the following:

$$\frac{1}{(1-p)^2} = 2kt[COOH]^2 + 1 = X_n^2$$

For a self-catalyzed system, the number average degree of polymerization (Xn) grows proportionally with $\sqrt{t}$.

External Catalyzed Polyesterification

The uncatalyzed reaction is rather slow, and a high X_n is not readily attained. In the presence of a catalyst, there is an acceleration of the rate, and the kinetic expression is altered to:

$$\frac{-d[\text{COOH}]}{dt} = k[\text{COOH}][\text{OH}]$$

which is kinetically first order in each functional group. Hence:

$$\frac{-d[\text{COOH}]}{dt} = k[\text{COOH}]^2$$

and integration gives finally:

$$\frac{1}{1-p} = 1 + [\text{COOH}]kt = X_n$$

For an externally catalyzed system, the number average degree of polymerization grows proportionally with t.

Molecular Weight Distribution in Linear Polymerization

The product of a polymerization is a mixture of polymer molecules of different molecular weights. For theoretical and practical reasons it is of interest to discuss the distribution of molecular weights in a polymerization. The molecular weight distribution (MWD) had been derived by Flory by a statistical approach based on the concept of equal reactivity of functional groups.

Probability

Step-growth polymerization is a random process so we can use statistics to calculate the probability of finding a chain with x-structural units ("x-mer") as a function of time or conversion.

$$x\,\text{AA} + x\,\text{BB} \rightarrow \text{AA-(BB-AA)}_{x-1}\text{-BB}$$
$$x\,\text{AB} \rightarrow \text{A-(B-A)}_{x-1}-\text{B}$$

Probability that an 'A' functional group has reacted

$$p^{x-1}$$

Probability of finding an 'A' unreacted

$$(1-p)$$

Combining the above two equations leads to.

$$P_x = (1-p)p^{x-1}$$

Where P_x is the probability of finding a chain that is x-units long and has an unreacted 'A'. As x increases the probability decreases.

Number Fraction Distribution

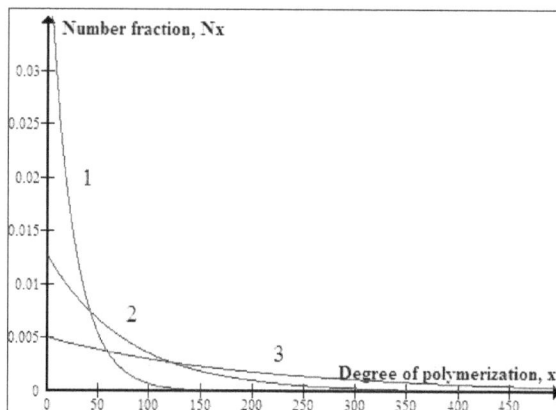

Number-fraction distribution curve for linear polymerization.
Plot 1, p=0.9600; plot 2, p=0.9875; plot 3, p=0.9950.

The number fraction distribution is the fraction of x-mers in any system and equals the probability of finding it in solution.

$$\frac{N_x}{N} = (1-p)p^{x-1}$$

where N is the total number of polymer molecules present in the reaction.

Weight Fraction Distribution

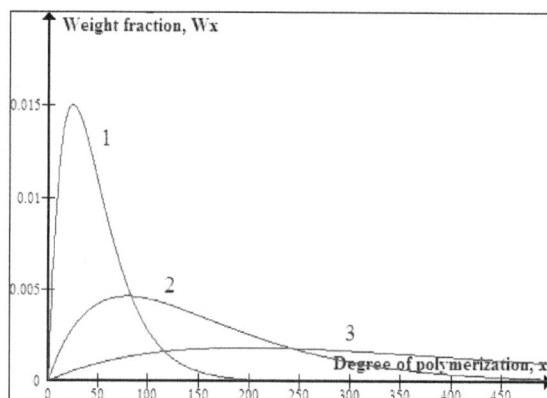

Weight fraction distribution plot for linear polymerization.
Plot 1, p=0.9600; plot 2, p=0.9875; plot 3, p=0.9950.

The weight fraction distribution is the fraction of x-mers in a system and the probability of finding them in terms of mass fraction:

$$\frac{W_x}{W_o} = \frac{xN_xM_o}{N_oM_o} = \frac{xN_x}{N_o} = x\frac{N_x}{N}\frac{N}{N_o}$$

Notes:

- M_o is the molar mass of the repeat unit,
- N_o is the initial number of monomer molecules,
- N is the number of unreacted functional groups.

Substituting from the Carothers equation:

$$X_n = \frac{1}{1-p} = \frac{N_o}{N}$$

We can now obtain:

$$\frac{W_x}{W_o} = x(1-p)^2 p^{x-1}$$

PDI

The polydispersity index (PDI), is a measure of the distribution of molecular mass in a given polymer sample:

$$PDI = \frac{M_w}{M_n}$$

However, for step-growth polymerization the Carothers equation can be used to substitute and rearrange this formula into the following:

$$PDI = 1 + p$$

Therefore, in step-growth when p=1, then the PDI=2.

Molecular Weight Control in Linear Polymerization

Need for Stoichiometric Control

There are two important aspects with regard to the control of molecular weight in polymerization. In the synthesis of polymers, one is usually interested in obtaining a product of very specific molecular weight, since the properties of the polymer will usually be highly dependent on molecular weight. Molecular weights higher or lower than the desired weight are equally undesirable. Since the degree of polymerization is a function of reaction time, the desired molecular weight can be obtained by quenching the reaction at the appropriate time. However, the polymer obtained in this manner is unstable in that it leads to changes in molecular weight because the ends of the polymer molecule contain functional groups that can react further with each other.

This situation is avoided by adjusting the concentrations of the two monomers so that they are slightly nonstoichiometric. One of the reactants is present in slight excess. The polymerization then proceeds to a point at which one reactant is completely used up and all the chain ends possess the same functional group of the group that is in excess. Further polymerization is not possible, and the polymer is stable to subsequent molecular weight changes.

Another method of achieving the desired molecular weight is by addition of a small amount of monofunctional monomer, a monomer with only one functional group. The monofunctional monomer, often referred to as a chain stopper, controls and limits the polymerization of bifunctional monomers because the growing polymer yields chain ends devoid of functional groups and therefore incapable of further reaction.

Quantitative Aspects

To properly control the polymer molecular weight, the stoichiometric imbalance of the bifunctional monomer or the monofunctional monomer must be precisely adjusted. If the nonstoichiometric imbalance is too large, the polymer molecular weight will be too low. It is important to understand the quantitative effect of the stoichiometric imbalance of reactants on the molecular weight. Also, this is necessary in order to know the quantitative effect of any reactive impurities that may be present in the reaction mixture either initially or that are formed by undesirable side reactions. Impurities with A or B functional groups may drastically lower the polymer molecular weight unless their presence is quantitatively taken into account.

More usefully, a precisely controlled stoichiometric imbalance of the reactants in the mixture can provide the desired result. For example, an excess of diamine over an acid chloride would eventually produce a polyamide with two amine end groups incapable of further growth when the acid chloride was totally consumed. This can be expressed in an extension of the Carothers equation as:

$$X_n = \frac{(1+r)}{(1+r-2rp)}$$

where r is the ratio of the number of molecules of the reactants:

$$r = \frac{N_{AA}}{N_{BB}}$$ were N_{BB} is the molecule in excess.

The equation above can also be used for a monofunctional additive which is the following:

$$r = \frac{N_{AA}}{(N_{BB} + 2N_B)}$$

where N_B is the number of monofunction molecules added. The coefficient of 2 in front of N_B is require since one B molecule has the same quantitative effect as one excess B-B molecule.

Multi-chain Polymerization

A monomer with functionality 3 has 3 functional groups which participate in the polymerization. This will introduce branching in a polymer and may ultimately form a cross-linked macrostructure. The point at which this three-dimensional 3D network is formed is known as the gel point, signaled by an abrupt change in viscosity.

A more general functionality factor f_{av} is defined for multi-chain polymerization, as the average number of functional groups present per monomer unit. For a system containing N_0 molecules initially and equivalent numbers of two function groups A and B, the total number of functional groups is $N_0 f_{av}$:

$$f_{av} = \frac{\sum N_i \cdot f_i}{\sum N_i}$$

And the modified Carothers equation is:

$$x_n = \frac{2}{2 - pf_{av}}, \text{ where p equals to } \frac{2(N_0 - N)}{N_0 \cdot f_{av}}$$

Advances in Step-growth Polymers

The driving force in designing new polymers is the prospect of replacing other materials of construction, especially metals, by using lightweight and heat-resistant polymers. The advantages of lightweight polymers include: high strength, solvent and chemical resistance, contributing to a variety of potential uses, such as electrical and engine parts on automotive and aircraft components, coatings on cookware, coating and circuit boards for electronic and microelectronic devices, etc. Polymer chains based on aromatic rings are desirable due to high bond strengths and rigid polymer chains. High molecular weight and crosslinking are desirable for the same reason. Strong dipole-dipole, hydrogen bond interactions and crystallinity also improve heat resistance. To obtain desired mechanical strength, sufficiently high molecular weights are necessary, however, decreased solubility is a problem. One approach to solve this problem is to introduce of some flexibilizing linkages, such as isopropylidene, C=O, and SO_2 into the rigid polymer chain by using an appropriate monomer or comonomer. Another approach involves the synthesis of reactive telechelic oligomers containing functional end groups capable of reacting with each other, polymerization of the oligomer gives higher molecular weight, referred to as chain extension.

Aromatic Polyether

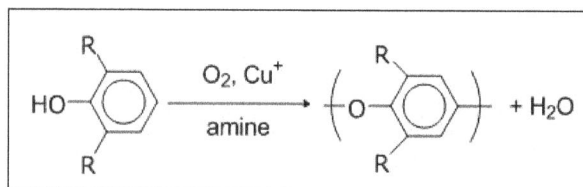

The oxidative coupling polymerization of many 2,6-disubstituted phenols using a catalytic complex of a cuprous salt and amine form aromatic polyethers, commercially referred to as poly(p-phenylene oxide) or PPO. Neat PPO has little commercial uses due to its high melt viscosity. Its available products are blends of PPO with high-impact polystyrene (HIPS).

Polyethersulfone

Polyethersulfone (PES) is also referred to as polyetherketone, polysulfone. It is synthesized by nucleophilic aromatic substitution between aromatic dihalides and bisphenolate salts. Polyethersulfones are partially crystalline, highly resistant to a wide range of aqueous and organic environment. They are rated for continuous service at temperatures of 240-280 °C. The polyketones are finding applications in areas like automotive, aerospace, electrical-electronic cable insulation.

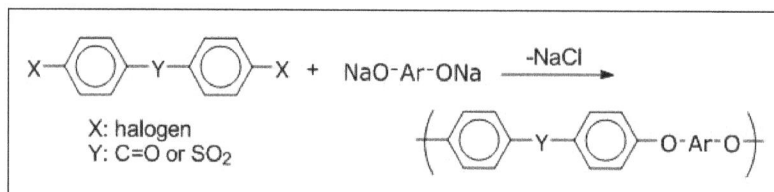

Aromatic Polysulfides

Poly(p-phenylene sulfide) (PPS) is synthesized by the reaction of sodium sulfide with p-dichlorobenzene in a polar solvent such as 1-methyl-2-pyrrolidinone (NMP). It is inherently flame-resistant and stable toward organic and aqueous conditions; however, it is somewhat susceptible to oxidants. Applications of PPS include automotive, microwave oven component, coating for cookware when blend with fluorocarbon polymers and protective coatings for valves, pipes, electromotive cells, etc.

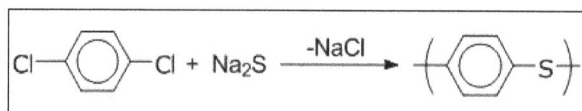

Aromatic Polyimide

Aromatic polyimides are synthesized by the reaction of dianhydrides with diamines, for example, pyromellitic anhydride with p-phenylenediamine. It can also be accomplished using diisocyanates in place of diamines. Solubility considerations sometimes suggest use of the half acid-half ester of the dianhydride, instead of the dianhydride itself. Polymerization is accomplished by a two-stage process due to the insolubility of polyimides. The first stage forms a soluble and fusible high-molecular-weight poly(amic acid) in a polar aprotic solvent such as NMP or N,N-dimethylacetamide. The poly(amic aicd) can then be processed into the desired physical form of the final plymer product (e.g., film, fiber, laminate, coating) which is insoluble and infusible.

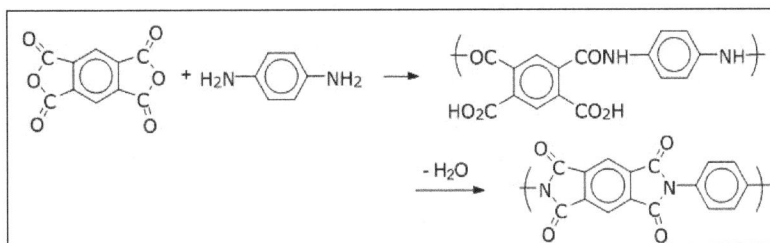

Telechelic Oligomer Approach

Telechelic oligomer approach applies the usual polymerization manner except that one includes a monofunctional reactant to stop reaction at the oligomer stage, generally in the 50-3000 molecular weight. The monofunctional reactant not only limits polymerization but end-caps the oligomer with functional groups capable of subsequent reaction to achieve curing of the oligomer. Functional groups like alkyne, norbornene, maleimide, nitrite, and cyanate have been used for this purpose. Maleimide and norbornene end-capped oligomers can be cured by heating. Alkyne, nitrile, and cyanate end-capped oligomers can undergo cyclotrimerization yielding aromatic structures.

Nitroxide-mediated Radical Polymerization

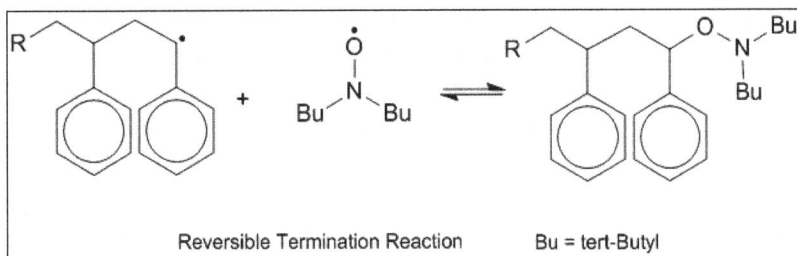

Reversible Termination Reaction   Bu = tert-Butyl

Nitroxide-mediated radical polymerization is a method of radical polymerization that makes use of an nitroxide initiator to generate polymers with well controlled stereochemistry and a very low polydispersity index. It is a type of reversible-deactivation radical polymerization.

Alkoxyamine Initiators

The initiating materials for nitroxide-mediated radical polymerization (NMP) are a family of compounds referred to as alkoxyamines. An alkoxyamine can essentially be viewed as an alcohol bound to a secondary amine by an N-O single bond. The utility of this functional group is that under certain conditions, homolysis of the C-O bond can occur, yielding a stable radical in the form of a 2-center 3-electron N-O system and a carbon radical which serves as an initiator for radical polymerization. For the purposes of NMP, the R groups attached to the nitrogen are always bulky, sterically hindering groups and the R group in the O- position forms a stable radical, generally is benzylic for polymerization to occur successfully. NMP allows for excellent control of chain length and structure, as well as a relative lack of true termination that allows polymerization to continue as long as there is available monomer. Because of this it is said to be "living".

Persistent Radical Effect

The living nature of NMP is due to the persistent radical effect (PRE). The PRE is a phenomenon observable in some radical systems which leads to the highly favored formation of one product to the near exclusion of other radical couplings due to one of the radical species being particularly stable, existing in greater and greater concentrations as the reaction progresses while the other one is transient, reacting quickly with either itself in a termination step or with the persistent radical to form a desired product. As time goes on, a higher concentration of the persistent radical is present, which couples reversibly with itself, meaning that any of the transient radical still present tends to couple with the persistent radical rather than itself due to greater availability. This leads to a greater proportion of cross-coupling than self-coupling in radical species.

In the case of a nitroxide-mediated polymerization reaction, the persistent radical is the nitroxide species and the transient radical is always the carbon radical. This leads to repeated coupling of the nitroxide to the growing end of the polymer chain, which would ordinarily be considered a termination step, but is in this case reversible. Because of the high rate of coupling of the nitroxide to the growing chain end, there is little coupling of two active growing chains, which would be an irreversible terminating step limiting the chain length. The nitroxide binds and unbinds to the growing chain, protecting it from termination steps. This ensures that any available monomer can be easily scavenged by active chains. Because this polymerization process does not naturally self-terminate, this polymerization process is described as "living," as the chains continue to grow under suitable reaction conditions whenever there is reactive monomer to "feed" them. Because of the PRE, it can be assumed that at any given time, almost all of the growing chains are "capped" by a mediating nitroxide, meaning that they dissociate and grow at very similar rates, creating a largely uniform chain length and structure.

Nitroxide Stability

Nitroxide radicals are effective mediators of well-controlled radical polymerization because they are quite stable, allowing them to act as persistent radicals in a reaction mixture. This stability is a result of their unique structure. In most diagrams, the radical is depicted on the oxygen, but another resonance structure exists which is more helpful in explaining their stability in which the radical is on the nitrogen, which has a double bond to the oxygen. In addition to this resonance stability, nitroxides used in NMRP always contain bulky, sterically hindering groups in the R1 and R2 positions. The significant steric bulk of these substituents entirely prevents radical coupling in the N-centered resonance form while significantly reducing it in the O-centered form. These bulky groups contribute stability, but only if there is no resonance provided by allyl or aromatic groups α to the N. These result in decreased stability of the nitroxide, presumably because they offer less sterically hindered sites for radical coupling to take place. The resulting inactivity of the radical makes hemolytic cleavage of the alkoxyamine quite fast in more sterically hindered species.

Nitroxide Choice

The choice of a specific nitroxide species to use has a large effect on the efficacy of an attempted polymerization. An effective polymerization (fast rate of chain growth, consistent chain length) results from a nitroxide with a fast C-O homolysis and relatively few side reactions. A more polar solvent lends itself better to C-O homolysis, so polar solvents which cannot bind to a labile nitroxide are the most effective for NMP. It is generally agreed that the structural factor that has the greatest effect on the ability of a nitroxide to mediate a radical polymerization is steric bulk. Generally speaking, greater steric bulk on the nitroxide leads to greater strain on the alkoxyamine, leading to the most easily broken bond, the C-O single bond, cleaving homolytically.

Ring Size

In the case of cyclic nitroxides, five-membered ring systems have been shown to cleave more slowly than six-membered rings and acyclic nitroxides with t-butyl moieties as their R groups cleaved fastest of all. This difference in the rate of cleavage was determined to result not from a difference in C-O bond lengths, but in the difference of C-O-N bond angle in the alkoxyamine. The smaller the bond angle the greater the steric interaction between the nitroxide and the alkyl fragment and the more easily the initiator species broke apart.

Steric Bulk

The efficiency of polymerization increases more and more with increased steric bulk of the nitroxide up to a point. TEMPO ((2,2,6,6-Tetramethylpiperidin-1-yl)oxyl) is capable of inducing the polymerization of styrene and styrene derivatives fairly easily, but is not sufficiently labile to induce polymerization of butyl acrylate under most conditions. TEMPO derivatives with even bulkier groups at the positions α to N have a rate of homolysis great enough to induce NMP of butyl acrylate, and the bulkier the α groups, the faster polymerization occurs. This indicates that the steric bulk of the nitroxide fragment can be a good indicator of the strength of an alkoxyamine initiator, at least up to a point. The equilibrium of its homolysis and reformation favors the radical form to the extent that recombination to reform an alkoxyamine over the course of NMP occurs too slowly to maintain control of chain length.

Preparation Methods

Because TEMPO, which is commercially available, is a sufficient nitroxide mediator for the synthesis of polystyrene derivatives, the preparation of alkoxyamine initiators for NMP of copolymers is in many cases a matter of attaching a nitroxide group (TEMPO) to a specifically synthesized alkyl fragment. Several methods have been reported to achieve this transformation.

Jacobsen's Catalyst

Jacobsen's catalyst is a manganese-based catalyst commonly used for the stereoselective epoxidation of alkenes. This epoxidation proceeds by a radical addition mechanism, which can be taken advantage of by introducing the radical TEMPO group into the reaction mixture. After treatment with a mild reducing agent such as sodium borohydride, this yields the product of a Markovnikov addition of nitroxide to the alkene. Jacobsen's catalyst is fairly mild, and a wide variety of functionalities on the alkene substrate can be tolerated. Practical yields are not necessarily as high as those reported by Dao et al, however.

Hydrazine

An alternative method is to react a substrate with a C-Br bond at the desired location of the nitroxide with hydrazine, generating an alkyl substituted hydrazine which is then exposed to a nitroxide radical and a mild oxidating agent such as lead dioxide. This generates a carbon-centered radical which couples with the nitroxide to generate the desired alkoxyamine. This method has the disadvantage of being relatively inefficient for some species, as well as the inherent danger of having to work with extremely toxic hydrazine and the inconvenience of having to run reactions in inert atmosphere.

Treatment of Aldehydes with Hydrogen Peroxide

Yet another published alkoxyamine synthesis involves treatment of aldehydes with hydrogen peroxide, which adds to the carbonyl group. The resulting species rearranges in situ in the presence of CuCl forming formic acid and the desired alkyl radical, which couples with tempo to produce the target alkoxyamine. The reaction appears to give fairly good yields and tolerates a variety of functional groups in the alkyl chain.

Electrophilic Bromination and Nucleophilic Attack

A synthesis has been described by Moon and Kang consisting of a one-electron reduction of a nitroxide radical in metallic sodium to yield a nucleophilic nitroxide. The nitroxide nucleophile is then added to an appropriate alkyl bromide, yielding the alkoxyamine by a simple SN2 reaction. This technique has the advantage of requiring only the appropriate alkyl bromide to be synthesized without requiring inconvenient reaction conditions and extremely hazardous reagents like Braso et al.'s method.

Ring-opening Polymerization

In polymer chemistry, ring-opening polymerization (ROP) is a form of chain-growth polymerization, in which the terminus of a polymer chain attacks cyclic monomers to form a longer polymer. The reactive center can be radical, anionic or cationic. Some cyclic monomers such as norbornene or cyclooctadiene can be polymerized to high molecular weight polymers by using metal catalysts. ROP is a versatile method for the synthesis of biopolymers.

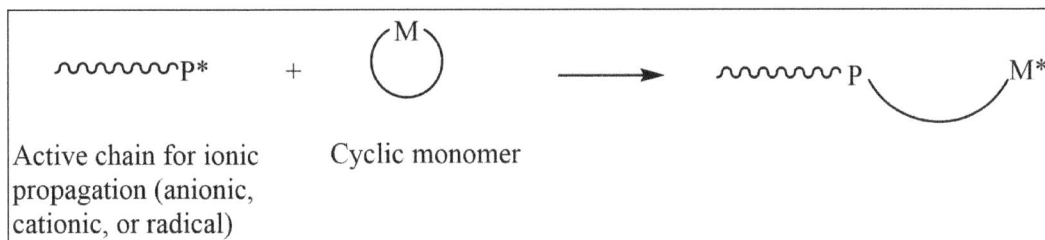

Active chain for ionic propagation (anionic, cationic, or radical)

Cyclic monomer

General scheme ionic propagation. Propagating center can be radical, cationic or anionic.

Ring-opening of cyclic monomers is often driven by the relief of bond-angle strain. Thus, as is the case for other types of polymerization, the enthalpy change in ring-opening is negative.

Monomers

Cyclic monomers that are amenable to ROP include epoxides, cyclic trisiloxanes, some lactones, lactides, cyclic carbonates, and amino acid N-carboxyanhydrides.. Many strained cycloalkenes, e.g norbornene, are suitable monomers via ring-opening metathesis polymerization.

Mechanisms

Ring-opening polymerization can proceed via radical, anionic, or cationic polymerization as described below. Additionally, radical ROP is useful in producing polymers with functional groups incorporated in the backbone chain that cannot otherwise be synthesized via conventional chain-growth polymerization of vinyl monomers. For instance, radical ROP can produce polymers with ethers, esters, amides, and carbonates as functional groups along the main chain.

Anionic Ring-opening Polymerization (AROP)

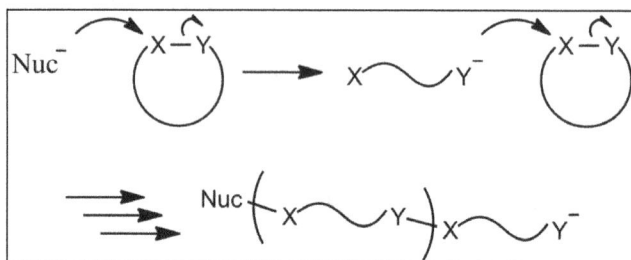

The general mechanism for anionic ring-opening polymerization. Polarized functional group is represented by X-Y, where the atom X (usually a carbon atom) becomes electron deficient due to the highly electron-withdrawing nature of Y (usually an oxygen, nitrogen, sulfur, etc.). The nucleophile will attack atom X, thus releasing Y-. The newly formed nucleophile will then attack the atom X in another monomer molecule, and the sequence would repeat until the polymer is formed.

Anionic ring-opening polymerizations (AROP) are involve nucleophilic reagents as initiators. Monomers with a three-member ring structure - such as epoxide, aziridine, and episulfide - undergo anionic ROP. A typical example of anionic ROP is that of ε-caprolactone, initiated by an alkoxide.

Cationic Ring-opening Polymerization

Cationic initiators and intermediates characterize cationic ring-opening polymerization (CROP). Examples of cyclic monomers that polymerize through this mechanism include lactones, lactams, amines, and ethers. CROP proceeds through an S_N1 or S_N2 propagation, chain-growth process. The mechanism is affected by the stability of the resulting cationic species. For example, if the atom bearing the positive charge is stabilized by electron-donating groups, polymerization will proceed by the S_N1 mechanism. The cationic species is a heteroatom and the chain grows by the addition of cyclic monomers thereby opening the ring system.

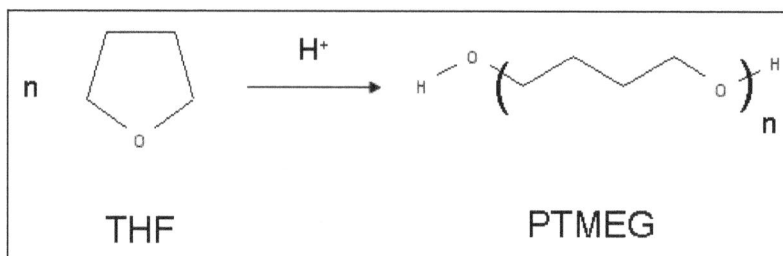

Synthesis of Spandex.

The monomers can be activated by Bronsted acids, carbenium ions, onium ions, and metal cations.

CROP can be a living polymerization and can be terminated by nucleophilic reagents such as phenoxy anions, phosphines, or polyanions. When the amount of monomers becomes depleted, termination can occur intra or intermolecularly. The active end can "backbite" the chain, forming a macrocycle. Alkyl chain transfer is also possible, where the active end is quenched by transferring an alkyl chain to another polymer.

Ring-opening Metathesis Polymerization

Ring-opening metathesis polymerization (ROMP) produces unsaturated polymers from cycloalkenes or bicycloalkenes. It requires organometallic catalysts.

The mechanism for ROMP follows similar pathways as olefin metathesis. The initiation process involves the coordination of the cycloalkene monomer to the metal alkylidene complex, followed by a [2+2] type cycloaddition to form the metallacyclobutane intermediate that cycloreverts to form a new alkylidene species.

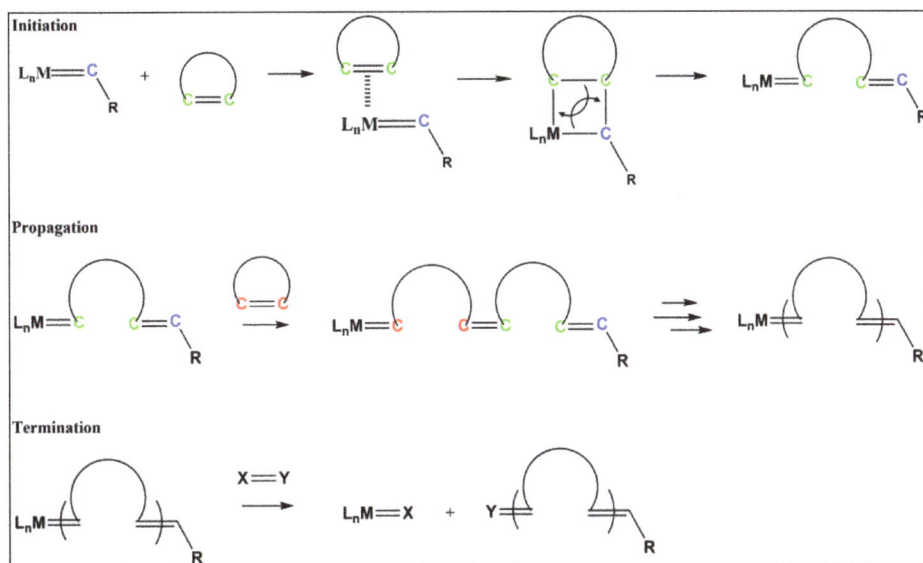

General scheme of the mechanism for ROMP.

Commercially relevant unsaturated polymers synthesized by ROMP include Norsorex (polynorbornene), Vestenamer (polycyclooctene), and Metton (polycyclopentadiene).

Thermodynamics

The formal thermodynamic criterion of a given monomer polymerizability is related to a sign of the free enthalpy (Gibbs free energy) of polymerization:

$$\Delta G_p(xy) = \Delta H_p(xy) - T\Delta S_p(xy)$$

where x and y indicate monomer and polymer states, respectively (x and/or y = l (liquid), g (gaseous), c (amorphous solid), c' (crystalline solid), s (solution)), $\Delta H_p(xy)$ and $\Delta Sp(xy)$ are the corresponding enthalpy (SI unit: joule per kelvin) and entropy (SI unit: joule) of polymerization, and T is the absolute temperature (SI unit: kelvin). The free enthalpy of polymerization (ΔG_p) may be expressed as a sum of standard enthalpy of polymerization (ΔG_p°) and a term related to instantaneous monomer molecules and growing macromolecules concentrations:

$$\Delta G_p = \Delta G_p^\circ + RT \ln \frac{[\ldots - (m)_{i+1} m^*]}{[M][\ldots - (m)_i m^*]}$$

where R is the gas constant, M is the monomer, $(m)_i$ is the monomer in an initial state, and m^* is the active monomer. Following Flory–Huggins solution theory that the reactivity of an active center, located at a macromolecule of a sufficiently long macromolecular chain, does not depend on its degree of polymerization (DPi), and taking in to account that $\Delta G_p^\circ = \Delta H_p^\circ - T\Delta S_p^\circ$ (where ΔH_p° and ΔS_p° indicate a standard polymerization enthalpy and entropy, respectively), we obtain:

$$\Delta G_p = \Delta H_p^\circ - T(\Delta S_p^\circ + R \ln[M])$$

At equilibrium ($\Delta G_p = 0$), when polymerization is complete the monomer concentration ($[M]_{eq}$) assumes a value determined by standard polymerization parameters (ΔH_p° and ΔS_p°) and polymerization temperature:

$$[M]_{eq} = e^{\frac{\Delta H_p^\circ}{RT} - \frac{\Delta S_p^\circ}{R}}$$

$$\ln \frac{DP_n}{DP_n - 1}[M]_{eq} = \frac{\Delta H_p^\circ}{RT} - \frac{\Delta S_p^\circ}{R}$$

$$[M]_{eq} = \frac{DP_n - 1}{DP_n} e^{\frac{\Delta H_p^\circ}{RT} - \frac{\Delta S_p^\circ}{R}}$$

Polymerzation is possible only when $[M]_0 > [M]_{eq}$. Eventually, at or above the so-called ceiling temperature (T_c), at which $[M]_{eq} = [M]_0$, formation of the high polymer does not occur.

$$T_c = \frac{\Delta H_p^\circ}{\Delta S_p^\circ + R \ln[M]_0}; (\Delta H_p^\circ < 0, \Delta S_p^\circ < 0)$$

$$T_f = \frac{\Delta H_p^\circ}{\Delta S_p^\circ + R \ln[M]_0}; (\Delta H_p^\circ > 0, \Delta S_p^\circ > 0)$$

For example, tetrahydrofuran (THF) cannot be polymerized above T_c = 84 °C, nor cyclo-octasulfur (S_8) below $T_f = 159$ °C. However, for many monomers, T_c and T_f, for polymerization in the bulk, are well above or below the operable polymerization temperatures, respectively. The polymerization of a majority of monomers is accompanied by an entropy decrease, due mostly to the loss in the translational degrees of freedom. In this situation, polymerization is thermodynamically allowed only when the enthalpic contribution into ΔG_p prevails (thus, when $\Delta H_p° < 0$ and $\Delta S_p° < 0$, the inequality $|\Delta H_p| > -T\Delta S_p$ is required). Therefore, the higher the ring strain, the lower the resulting monomer concentration at equilibrium.

Suspension Polymerization

Suspension polymerization, sometimes called bead, pearl or granular polymerization, is one of the most widely used polymerization techniques. It is essentially a water or solvent cooled bulk polymerization, though water/solvent soluble initiators may be present that could alter the reaction kinetic.

Practically all common themoplastic polymers can be made by this method. This includes all high volume resins such as methyl methacrylate, vinyl chloride, vinyl acetate, stryrene and some gaseous monomers such as ethylene, propylene and formaldehyde.

Suspension polymerization has several advantages over other polymerization techniques; since water is usually the continuous phase, it acts as a very effective heat-tranfer medium which is very ecconomical and more environmental friendly than the solvents employed in solution polymerization. Furthermore, temperature and viscosity control is fairly easy. Compared to emulsion polymerization, purification and processing of the polymer is much easier since very little catalyst is used and the final product is a 100% solid resin.

In general, a suspension polymerization system consists of a dispersing medium, monomer(s), stabilizing agents and a monomer soluble initiator. Water is almost in all cases the continuous phase. For the monomer to be dispersed in the water, it has to be fairly insoluble. If the monomer is not sufficient insoluble, prepolymers or partially polymerized monomers (oligomers) can be used that are either insoluble or have a much lower solubility in the water than the monomers. This will also increase the particle size.

Suspension Polymerization Vessel

The initiators employed for suspension polymerization are mostly peroxides, and in some cases, azo compounds (AIBN). Typical initiators are benzoyl, t-butyl, diacetyl and lauroyl peroxide. Typical stabilizers are surfactants such as sodium, potassium, or ammonium salts of fatty acids that lower the surface tension and *dispersing agents* such as polyelectrolytes and some inorganic salts that stabilize the particles by providing a surface charge (electro-static interaction or electro-steric hindrance). Common polyelectrolytes are alkali salts of poly(methyl acrylate) and poly(methyl methacrylate) and common salts are magnesium and calcium carbonate, calcium phosphate, and aluminum chloride. The most common stabilizers are water soluble, non-micelle-forming polymers such as methyl and ethyl cellulose, poly(vinyl alcohol), gums, alginates, casein, gelatins, and starches and (nano)particles such as talc, silicates, clays and bentonites. The surfactants and electrolytes stabilize the monomer droplets and polymers particles and reduce the viscosity, whereas the water soluble polymers and dispersed (nano)particles increase the viscosity and act as a protective coating. This prevents coalescence or agglomeration of the sticky polymer particles. Usually, very small amounts of stabilizing agents are sufficient to stabilize the monomer particles. In the case of surfactants, the concentration is usually not more than 0.01 % and for dispersing agent, it is around 0.1 % based on the water. Then the monomer or monomers blend is added so that the overall concentration is not more than 25 – 40 %.

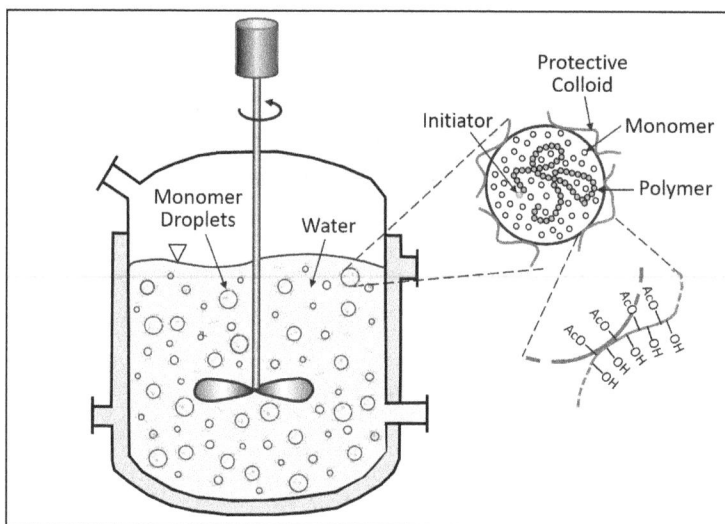

The polymerization is carried out in the small droplets of liquid monomer. It can be considered as a form of mass polymerization on a much smaller scale, that is, the reaction steps and the kinetics are the same as those in bulk polymerization and consist of initiation, propagation and termination. During the polymerization, the immiscible droplets slowly convert from a liquid to a sticky, viscose material and, when reaching a sufficient high molecular weight, form solid, rigid particles.

During the polymerization, the suspension has to be stirred and sufficiently stabilized or the particles agglomerate and form one big mass. The final product is usually not sticky if the glass transition temperature or melting point is above the granulate temperature. The final polymer particles have a size in the range 0.1 to 5 mm in diameter and are at least ten times larger than those produced during emulsion polymerisation.

Radical Polymerization

Free-radical polymerization (FRP) is a method of polymerization by which a polymer forms by the successive addition of free-radical building blocks. Free radicals can be formed by a number of different mechanisms, usually involving separate initiator molecules. Following its generation, the initiating free radical adds (nonradical) monomer units, thereby growing the polymer chain.

Free-radical polymerization is a key synthesis route for obtaining a wide variety of different polymers and material composites. The relatively non-specific nature of free-radical chemical interactions makes this one of the most versatile forms of polymerization available and allows facile reactions of polymeric free-radical chain ends and other chemicals or substrates. In 2001, 40 billion of the 110 billion pounds of polymers produced in the United States were produced by free-radical polymerization.

Free-radical polymerization is a type of chain-growth polymerization, along with anionic, cationic and coordination polymerization.

Initiation

Initiation is the first step of the polymerization process. During initiation, an active center is created from which a polymer chain is generated. Not all monomers are susceptible to all types of initiators. Radical initiation works best on the carbon–carbon double bond of vinyl monomers and the carbon–oxygen double bond in aldehydes and ketones. Initiation has two steps. In the first step, one or two radicals are created from the initiating molecules. In the second step, radicals are transferred from the initiator molecules to the monomer units present. Several choices are available for these initiators.

Types of Initiation and the Initiators

Thermal decomposition: The initiator is heated until a bond is homolytically cleaved, producing two radicals. This method is used most often with organic peroxides or azo compounds.

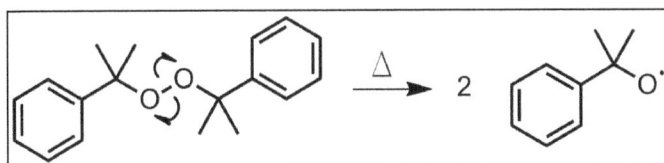

Thermal decomposition of dicumyl peroxide.

Photolysis: Radiation cleaves a bond homolytically, producing two radicals. This method is used most often with metal iodides, metal alkyls, and azo compounds.

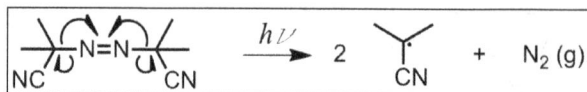

Photolysis of azoisobutylnitrile (AIBN).

Photoinitiation can also occur by bi-molecular H abstraction when the radical is in its lowest triplet excited state. An acceptable photoinitiator system should fulfill the following requirements:

- High absorptivity in the 300–400 nm range.

- Efficient generation of radicals capable of attacking the alkene double bond of vinyl monomers.

- Adequate solubility in the binder system (prepolymer + monomer).

- Should not impart yellowing or unpleasant odors to the cured material.

- The photoinitiator and any byproducts resulting from its use should be non-toxic.

Redox reactions: Reduction of hydrogen peroxide or an alkyl hydrogen peroxide by iron. Other reductants such as Cr^{2+}, V^{2+}, Ti^{3+}, Co^{2+}, and Cu^+ can be employed in place of ferrous ion in many instances.

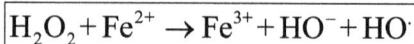

$$H_2O_2 + Fe^{2+} \rightarrow Fe^{3+} + HO^- + HO^{\cdot}$$

Redox reaction of hydrogen peroxide and iron.

Persulfates: The dissociation of a persulfate in the aqueous phase. This method is useful in emulsion polymerizations, in which the radical diffuses into a hydrophobic monomer-containing droplet.

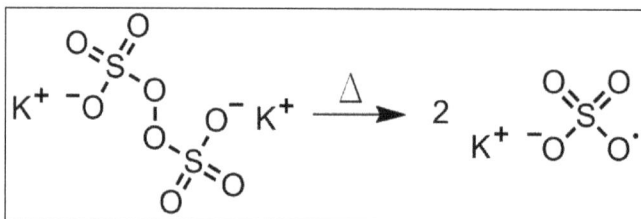

Thermal degradation of a persulfate.

Ionizing radiation: α-, β-, γ-, or x-rays cause ejection of an electron from the initiating species, followed by dissociation and electron capture to produce a radical.

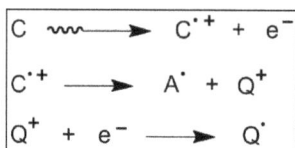

$$C \rightsquigarrow C^{\cdot +} + e^-$$
$$C^{\cdot +} \rightarrow A^{\cdot} + Q^+$$
$$Q^+ + e^- \rightarrow Q^{\cdot}$$

The three steps involved in ionizing radiation: ejection, dissociation, and electron-capture.

Electrochemical: Electrolysis of a solution containing both monomer and electrolyte. A monomer molecule will receive an electron at the cathode to become a radical anion, and a monomer molecule will give up an electron at the anode to form a radical cation. The radical ions then initiate free radical (and/or ionic) polymerization. This type of initiation of especially useful for coating metal surfaces with polymer films.

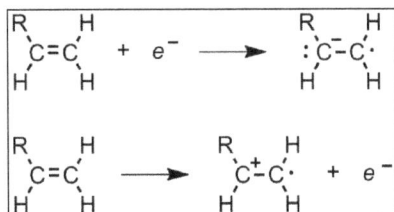

Formation of radical anion at the cathode; (bottom) formation of radical cation at the anode.

Plasma: A gaseous monomer is placed in an electric discharge at low pressures under conditions where a plasma (ionized gaseous molecules) is created. In some cases, the system is heated and/or placed in a radiofrequency field to assist in creating the plasma.

Sonication: High-intensity ultrasound at frequencies beyond the range of human hearing (16 kHz) can be applied to a monomer. Initiation results from the effects of cavitation (the formation and collapse of cavities in the liquid). The collapse of the cavities generates very high local temperatures and pressures. This results in the formation of excited electronic states, which in turn lead to bond breakage and radical formation.

Ternary initiators: A ternary initiator is the combination of several types of initiators into one initiating system. The types of initiators are chosen based on the properties they are known to induce in the polymers they produce. For example, poly(methyl methacrylate) has been synthesized by the ternary system benzoyl peroxide-3,6-bis(o-carboxybenzoyl)-N-isopropylcarbazole-di-η^5-indenylzicronium dichloride.

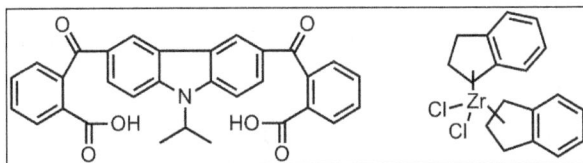

Benzoyl peroxide-3,6-bis(o-carboxybenzoyl)-N-isopropylcarbazole-di-η^5-indenylzicronium dichloride.

This type of initiating system contains a metallocene, an initiator, and a heteroaromatic diketo carboxylic acid. Metallocenes in combination with initiators accelerate polymerization of poly(methyl methacrylate) and produce a polymer with a narrower molecular weight distribution. The example shown here consists of indenylzirconium (a metallocene) and benzoyl peroxide (an initiator). Also, initiating systems containing heteroaromatic diketo carboxylic acids, such as 3,6-bis(o-carboxybenzoyl)-N-isopropylcarbazole in this example, are known to catalyze the decomposition of benzoyl peroxide. Initiating systems with this particular heteroaromatic diket carboxylic acid are also known to have effects on the microstructure of the polymer. The combination of all of these components—a metallocene, an initiator, and a heteroaromatic diketo carboxylic acid—yields a ternary initiating system that was shown to accelerate the polymerization and produce polymers with enhanced heat resistance and regular microstructure.

Initiator Efficiency

Due to side reactions and inefficient synthesis of the radical species, chain initiation is not 100%. The efficiency factor f is used to describe the effective radical concentration. The maximal value of f is 1, but typical values range from 0.3 to 0.8. The following is a list of reactions that decrease the efficiency of the initiator.

Primary recombination: Two radicals recombine before initiating a chain. This occurs within the solvent cage, meaning that no solvent has yet come between the new radicals.

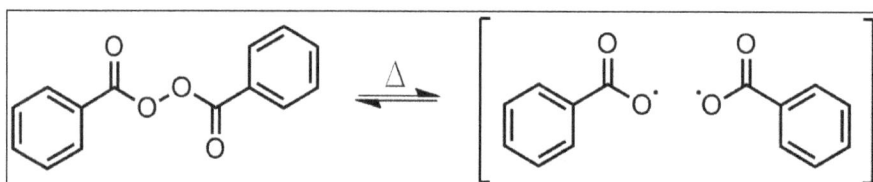

Primary recombination of BPO; brackets indicate that the reaction is happening within the solvent cage.

Other recombination pathways: Two radical initiators recombine before initiating a chain, but not in the solvent cage.

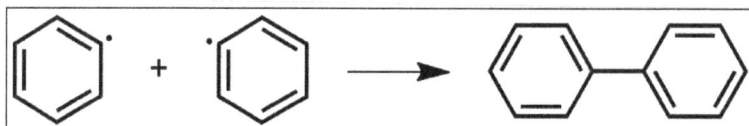

Recombination of phenyl radicals from the initiation of BPO outside the solvent cage.

Side reactions: One radical is produced instead of the three radicals that could be produced. Reaction of polymer chain R:

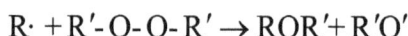

$$R\cdot + R'\text{-}O\text{-}O\text{-}R' \rightarrow ROR' + R'O'$$

Propagation

During polymerization, a polymer spends most of its time in increasing its chain length, or propagating. After the radical initiator is formed, it attacks a monomer. In an ethene monomer, one electron pair is held securely between the two carbons in a sigma bond. The other is more loosely held in a pi bond. The free radical uses one electron from the pi bond to form a more stable bond with the carbon atom. The other electron returns to the second carbon atom, turning the whole molecule into another radical. This begins the polymer chain. Figure shows how the orbitals of an ethylene monomer interact with a radical initiator.

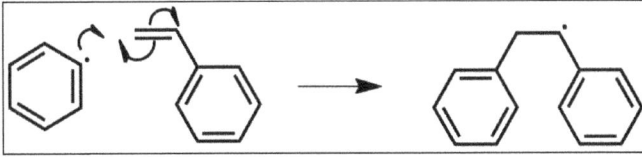

Phenyl initiator from benzoyl peroxide (BPO) attacks a styrene molecule to start the polymer chain.

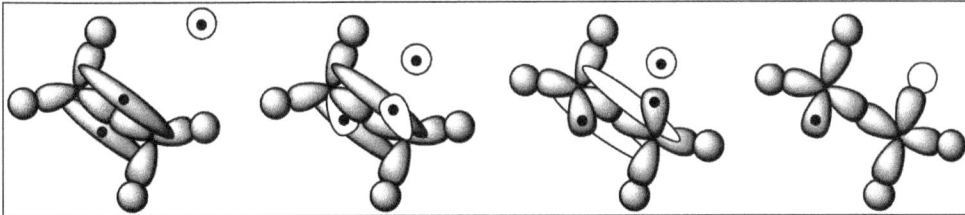

An orbital drawing of the initiator attack on ethylene molecule,
producing the start of the polyethylene chain.

Once a chain has been initiated, the chain propagates until there are no more monomers (living polymerization) or until termination occurs. There may be anywhere from
a few to thousands of propagation steps depending on several factors such as radical
and chain reactivity, the solvent, and temperature. The mechanism of chain propagation is as follows.

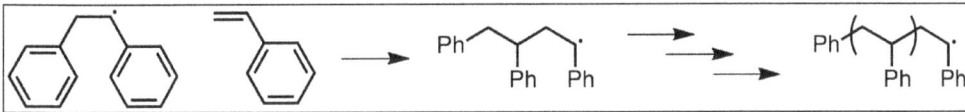

Propagation of polystyrene with a phenyl radical initiator.

Termination

Chain termination is inevitable in radical polymerization due to the high reactivity of radicals. Termination can occur by several different mechanisms. If longer chains are desired,
the initiator concentration should be kept low; otherwise, many shorter chains will result.

- Combination of two active chain ends: one or both of the following processes
  may occur.

  o Combination: Two chain ends simply couple together to form one long
     chain. One can determine if this mode of termination is occurring by mon-
     itoring the molecular weight of the propagating species: combination will
     result in doubling of molecular weight. Also, combination will result in a
     polymer that is C_2 symmetric about the point of the combination.

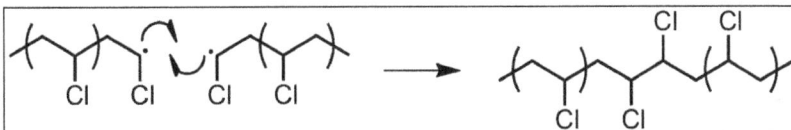

Termination by the combination of two poly(vinyl chloride) (PVC) polymers.

- o   Radical disproportionation: A hydrogen atom from one chain end is abstracted to another, producing a polymer with a terminal unsaturated group and a polymer with a terminal saturated group.

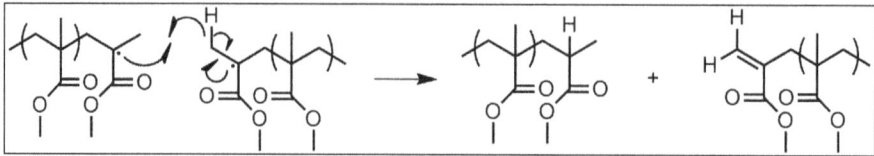

Termination by disproportionation of poly(methyl methacrylate).

- Combination of an active chain end with an initiator radical.

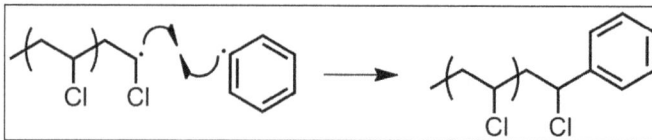

Termination of PVC by reaction with radical initiator.

- Interaction with impurities or inhibitors. Oxygen is the common inhibitor. The growing chain will react with molecular oxygen, producing an oxygen radical, which is much less reactive. This significantly slows down the rate of propagation.

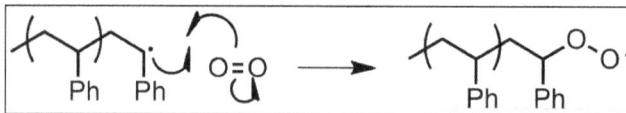

Inhibition of polystyrene propagation due to reaction of polymer with molecular oxygen.

Nitrobenzene, butylated hydroxyl toluene, and diphenyl picryl hydrazyl are a few other inhibitors. The latter is an especially effective inhibitor because of the resonance stabilization of the radical.

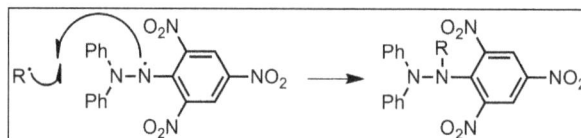

Inhibition of polymer chain, R, by DPPH.

Chain Transfer

Contrary to the other modes of termination, chain transfer results in the destruction of only one radical, but also the creation of another radical. Often, however, this newly created radical is not capable of further propagation. Similar to disproportionation, all chain transfer mechanisms also involve the abstraction of a hydrogen or other atom. There are several types of chain transfer mechanisms.

- To solvent: A hydrogen atom is abstracted from a solvent molecule, resulting in the formation of radical on the solvent molecules, which will not propagate further.

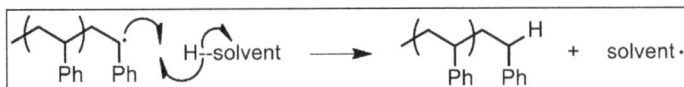
Chain transfer from polystyrene to solvent.

The effectiveness of chain transfer involving solvent molecules depends on the amount of solvent present (more solvent leads to greater probability of transfer), the strength of the bond involved in the abstraction step (weaker bond leads to greater probability of transfer), and the stability of the solvent radical that is formed (greater stability leads to greater probability of transfer). Halogens, except fluorine, are easily transferred.

- To monomer: A hydrogen atom is abstracted from a monomer. While this does create a radical on the affected monomer, resonance stabilization of this radical discourages further propagation.

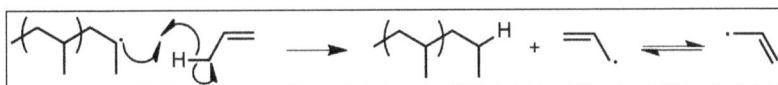
Chain transfer from polypropylene to monomer.

- To initiator: A polymer chain reacts with an initiator, which terminates that polymer chain, but creates a new radical initiator. This initiator can then begin new polymer chains. Therefore, contrary to the other forms of chain transfer, chain transfer to the initiator does allow for further propagation. Peroxide initiators are especially sensitive to chain transfer.

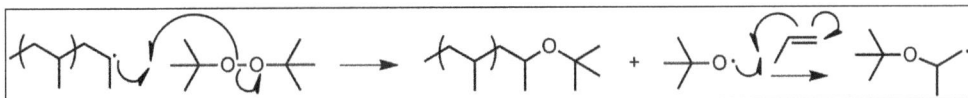
Chain transfer from polypropylene to di-t-butyl peroxide initiator.

- To polymer: the radical of a polymer chain abstracts a hydrogen atom from somewhere on another polymer chain. This terminates the growth of one polymer chain, but allows the other to branch and resume growing. This reaction step changes neither the number of polymer chains nor the number of monomers which have been polymerized, so that the number-average degree of polymerization is unaffected.

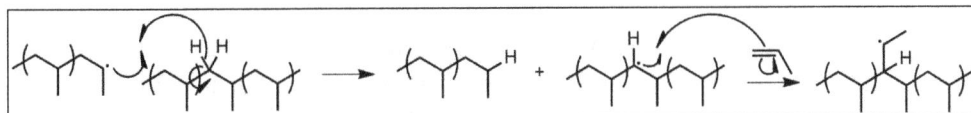
Chain transfer from polypropylene to backbone of another polypropylene.

Effects of chain transfer: The most obvious effect of chain transfer is a decrease in the polymer chain length. If the rate of transfer is much larger than the rate of propagation, then very small polymers are formed with chain lengths of 2-5 repeating units (telomerization). The Mayo equation estimates the influence of chain transfer on chain length (x_n): $\dfrac{1}{x_n} = \left(\dfrac{1}{x_n}\right)_0 + \dfrac{k_{tr}[solvent]}{k_p[monomer]}$. Where k_{tr} is the rate constant for chain transfer and k_p is the rate constant for propagation. The Mayo equation assumes that transfer to solvent is the major termination pathway.

Methods

There are four industrial methods of radical polymerization:

- Bulk polymerization: Reaction mixture contains only initiator and monomer, no solvent.

- Solution polymerization: Reaction mixture contains solvent, initiator, and monomer.

- Suspension polymerization: Reaction mixture contains an aqueous phase, water-insoluble monomer, and initiator soluble in the monomer droplets (both the monomer and the initiator are hydrophobic).

- Emulsion polymerization: Similar to suspension polymerization except that the initiator is soluble in the aqueous phase rather than in the monomer droplets (the monomer is hydrophobic, and the initiator is hydrophilic). An emulsifying agent is also needed.

Other methods of radical polymerization include the following:

- Template polymerization: In this process, polymer chains are allowed to grow along template macromolecules for the greater part of their lifetime. A well-chosen template can affect the rate of polymerization as well as the molar mass and microstructure of the daughter polymer. The molar mass of a daughter polymer can be up to 70 times greater than those of polymers produced in the absence of the template and can be higher in molar mass than the templates themselves. This is because of retardation of the termination for template-associated radicals and by hopping of a radical to the neighboring template after reaching the end of a template polymer.

- Plasma polymerization: The polymerization is initiated with plasma. A variety of organic molecules including alkenes, alkynes, and alkanes undergo polymerization to high molecular weight products under these conditions. The propagation mechanisms appear to involve both ionic and radical species. Plasma polymerization offers a potentially unique method of forming thin polymer films for uses such as thin-film capacitors, antireflection coatings, and various types of thin membranes.

- Sonication: The polymerization is initiated by high-intensity ultrasound. Polymerization to high molecular weight polymer is observed but the conversions are low (<15%). The polymerization is self-limiting because of the high viscosity produced even at low conversion. High viscosity hinders cavitation and radical production.

Reversible Deactivation Radical Polymerization

Also known as living radical polymerization, controlled radical polymerization, reversible deactivation radical polymerization (RDRP) relies on completely pure reactions, preventing termination caused by impurities. Because these polymerizations stop only when there is no more monomer, polymerization can continue upon the addition of more monomer. Block copolymers can be made this way. RDRP allows for control of molecular weight and dispersity. However, this is very difficult to achieve and instead a pseudo-living polymerization occurs with only partial control of molecular weight and dispersity. ATRP and RAFT are the main types of complete radical polymerization.

- Atom transfer radical polymerization (ATRP): Based on the formation of a carbon-carbon bond by atom transfer radical addition. This method, independently discovered in 1995 by Mitsuo Sawamoto and by Jin-Shan Wang and Krzysztof Matyjaszewski, requires reversible activation of a dormant species (such as an alkyl halide) and a transition metal halide catalyst (to activate dormant species).

- Reversible Addition-Fragmentation Chain Transfer Polymerization (RAFT): Requires a compound that can act as a reversible chain transfer agent, such as dithio compounds.

- Stable Free Radical Polymerization (SFRP): Used to synthesize linear or branched polymers with narrow molecular weight distributions and reactive end groups on each polymer chain. The process has also been used to create block co-polymers with unique properties. Conversion rates are about 100% using this process but require temperatures of about 135 °C. This process is most commonly used with acrylates, styrenes, and dienes. The reaction scheme in figure illustrates the SFRP process.

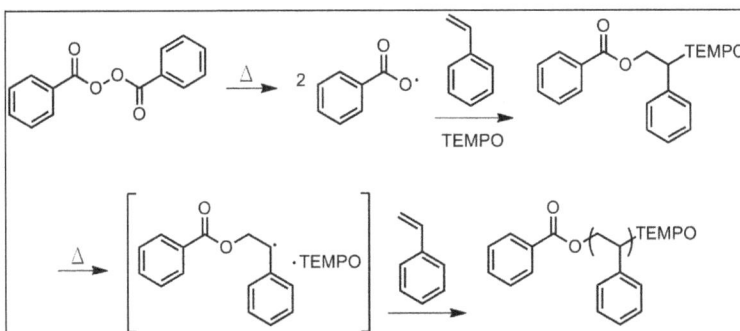

Reaction scheme for SFRP.

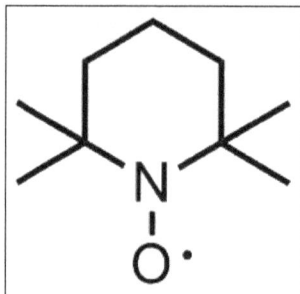

TEMPO molecule used to functionalize the chain ends.

Because the chain end is functionalized with the TEMPO molecule, premature termination by coupling is reduced. As with all living polymerizations, the polymer chain grows until all of the monomer is consumed.

Kinetics

In typical chain growth polymerizations, the reaction rates for initiation, propagation and termination can be described as follows:

$$v_i = d[M \cdot]/dt = 2k_d f[I]$$

$$v_p = k_p[M][M \cdot]$$

$$v_t = -d[M \cdot]/dt = 2k_t[M \cdot]^2$$

where f is the efficiency of the initiator and k_d, k_p, and k_t are the constants for initiator dissociation, chain propagation and termination, respectively. [I] [M] and [M·] are the concentrations of the initiator, monomer and the active growing chain.

Under the steady state approximation, the concentration of the active growing chains remains constant, i.e. the rates of initiation and of termination are equal. The concentration of active chain can be derived and expressed in terms of the other known species in the system:

$$[M \cdot] = \left(\frac{k_d[I]f}{k_t} \right)^{1/2}$$

In this case, the rate of chain propagation can be further described using a function of the initiator and monomer concentrations:

$$v_p = k_p \left(\frac{f k_d}{k_t} \right)^{1/2} [I]^{1/2}[M]$$

The kinetic chain length v is a measure of the average number of monomer units reacting with an active center during its lifetime and is related to the molecular weight through the mechanism of the termination. Without chain transfer, the kinetic chain length is only a function of propagation rate and initiation rate:

$$v = \frac{v_p}{v_i} = \frac{k_p[M][M\cdot]}{2fk_d[I]} = \frac{k_p[M]}{2(fk_dk_t[I])^{1/2}}$$

Assuming no chain transfer effect occurs in the reaction, the number average degree of polymerization P_n can be correlated with the kinetic chain length. In the case of termination by disproportionation, one polymer molecule is produced per every kinetic chain:

$$x_n = v$$

Termination by combination leads to one polymer molecule per two kinetic chains:

$$x_n = 2v$$

Any mixture of both these mechanisms can be described by using the value δ, the contribution of disproportionation to the overall termination process:

$$x_n = \frac{2}{1+\delta}v$$

If chain transfer is considered, the kinetic chain length is not affected by the transfer process because the growing free-radical center generated by the initiation step stays alive after any chain transfer event, although multiple polymer chains are produced. However, the number average degree of polymerization decreases as the chain transfers, since the growing chains are terminated by the chain transfer events. Taking into account the chain transfer reaction towards solvent S, initiator I, polymer P, and added chain transfer agent T. The equation of P_n will be modified as follows:

$$\frac{1}{x_n} = \frac{2k_{t,d} + k_{t,c}}{k_p^2[M]^2}v_p + C_M + C_S\frac{[S]}{[M]} + C_I\frac{[I]}{[M]} + C_P\frac{[P]}{[M]} + C_T\frac{[T]}{[M]}$$

It is usual to define chain transfer constants C for the different molecules:

$$C_M = \frac{k_{tr}^M}{k_p}, \; C_S = \frac{k_{tr}^S}{k_p}, \; C_I = \frac{k_{tr}^I}{k_p}, \; C_P = \frac{k_{tr}^P}{k_p}, \; C_T = \frac{k_{tr}^T}{k_p}$$

Thermodynamics

In chain growth polymerization, the position of the equilibrium between polymer and monomers can be determined by the thermodynamics of the polymerization. The Gibbs free energy (ΔG_p) of the polymerization is commonly used to quantify the tendency of a polymeric reaction. The polymerization will be favored if $\Delta G_p < 0$; if $\Delta G_p > 0$, the polymer will undergo depolymerization. According to the thermodynamic equation $\Delta G = \Delta H - T\Delta S$, a negative enthalpy and an increasing entropy will shift the equilibrium towards polymerization.

In general, the polymerization is an exothermic process, i.e. negative enthalpy change, since addition of a monomer to the growing polymer chain involves the conversion of π bonds into σ bonds, or a ring–opening reaction that releases the ring tension in a cyclic monomer. Meanwhile, during polymerization, a large amount of small molecules are associated, losing rotation and translational degrees of freedom. As a result, the entropy decreases in the system, $\Delta S_p < 0$ for nearly all polymerization processes. Since depolymerization is almost always entropically favored, the ΔH_p must then be sufficiently negative to compensate for the unfavorable entropic term. Only then will polymerization be thermodynamically favored by the resulting negative ΔG_p.

In practice, polymerization is favored at low temperatures: $T\Delta S_p$ is small. Depolymerization is favored at high temperatures: $T\Delta S_p$ is large. As the temperature increases, ΔG_p become less negative. At a certain temperature, the polymerization reaches equilibrium (rate of polymerization = rate of depolymerization). This temperature is called the ceiling temperature (T_c). $\Delta G_p = 0$.

Stereochemistry

The stereochemistry of polymerization is concerned with the difference in atom connectivity and spatial orientation in polymers that has the same chemical composition. Staudinger studied the stereoisomerism in chain polymerization of vinyl monomers in late 1920s, and it took another two decades for people to fully appreciate the idea that each of the propagation steps in the polymer growth could give rise to stereoisomerism. The major milestone in the stereochemistry was established by Ziegler and Natta and their coworkers in 1950s, as they developed metal based catalyst to synthesize stereoregular polymers. The reason why the stereochemistry of the polymer is of particular interest is because the physical behavior of a polymer depends not only on the general chemical composition but also on the more subtle differences in microstructure. Atactic polymers consist of a random arrangement of stereochemistry and are amorphous (noncrystalline), soft materials with lower physical strength. The corresponding isotactic (like substituents all on the same side) and syndiotactic (like substituents of alternate repeating units on the same side) polymers are usually obtained as highly crystalline materials. It is easier for the stereoregular polymers to pack into a crystal lattice since they are more ordered and the resulting crystallinity leads to higher physical strength and increased solvent and chemical resistance as well as differences in other properties that depend on crystallinity. The prime example of the industrial utility of stereoregular polymers is polypropene. Isotactic polypropene is a high-melting (165 °C), strong, crystalline polymer, which is used as both a plastic and fiber. Atactic polypropene is an amorphous material with an oily to waxy soft appearance that finds use in asphalt blends and formulations for lubricants, sealants, and adhesives, but the volumes are minuscule compared to that of isotactic polypropene.

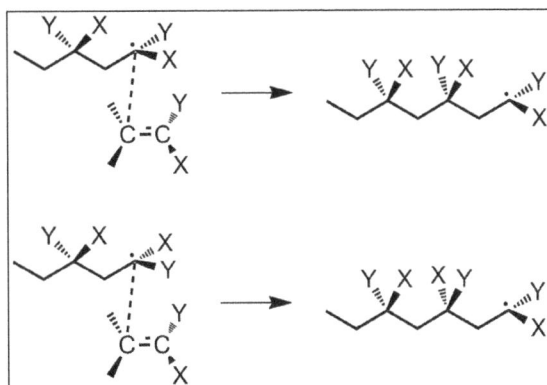

(Top) formation of isotactic polymer; (bottom) formation of syndiotactic polymer.

When a monomer adds to a radical chain end, there are two factors to consider regarding its stereochemistry: 1) the interaction between the terminal chain carbon and the approaching monomer molecule and 2) the configuration of the penultimate repeating unit in the polymer chain. The terminal carbon atom has sp^2 hybridization and is planar. Consider the polymerization of the monomer $CH_2=CXY$. There are two ways that a monomer molecule can approach the terminal carbon: the mirror approach (with like substituents on the same side) or the non-mirror approach (like substituents on opposite sides). If free rotation does not occur before the next monomer adds, the mirror approach will always lead to an isotactic polymer and the non-mirror approach will always lead to a syndiotactic polymer.

However, if interactions between the substituents of the penultimate repeating unit and the terminal carbon atom are significant, then conformational factors could cause the monomer to add to the polymer in a way that minimizes steric or electrostatic interaction.

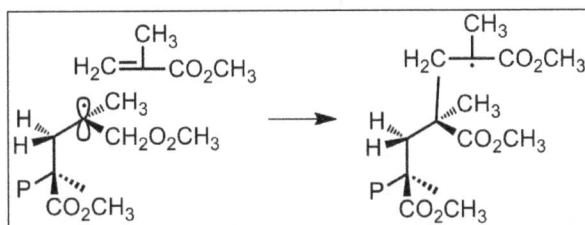

Penultimate unit interactions cause monomer to add in a way that minimizes steric hindrance between substituent groups. P represents polymer chain.

Reactivity

Traditionally, the reactivity of monomers and radicals are assessed by the means of copolymerization data. The Q–e scheme, the most widely used tool for the semi-quantitative prediction of monomer reactivity ratios, was first proposed by Alfrey and Price in 1947. The scheme takes into account the intrinsic thermodynamic stability and polar effects in the transition state. A given radical M_i^o and a monomer M_j are considered to have intrinsic reactivities P_i and Q_j, respectively. The polar effects in the transition state, the supposed permanent electric charge carried by that entity (radical or molecule), is quantified by the factor e, which is a constant for a given monomer, and has the same value for the radical derived from that specific monomer. For addition of monomer 2 to a growing polymer chain whose active end is the radical of monomer 1, the rate constant, k_{12}, is postulated to be related to the four relevant reactivity parameters by:

$$k_{12} = P_1 Q_2 \exp(-e_1 e_2)$$

The monomer reactivity ratio for the addition of monomers 1 and 2 to this chain is given by:

$$r_1 = \frac{k_{11}}{k_{12}} = \frac{Q_1}{Q_2} \exp(-e_1(e_1 - e_2))$$

For the copolymerization of a given pair of monomers, the two experimental reactivity ratios r_1 and r_2 permit the evaluation of (Q_1/Q_2) and $(e_1 - e_2)$. Values for each monomer can then be assigned relative to a reference monomer, usually chosen as styrene with the arbitrary values $Q = 1.0$ and $e = -0.8$.

Applications

Free radical polymerization has found applications including the manufacture of polystyrene, thermoplastic block copolymer elastomers, cardiovascular stents, chemical surfactants and lubricants. Block copolymers are used for a wide variety of applications including adhesives, footwear and toys.

Free radical polymerization has uses in research as well, such as in the functionalization of carbon nanotubes. CNTs intrinsic electronic properties lead them to form large aggregates in solution, precluding useful applications. Adding small chemical groups to the walls of CNT can eliminate this propensity and tune the response to the surrounding environment. The use of polymers instead of smaller molecules can modify CNT properties (and conversely, nanotubes can modify polymer mechanical and electronic properties). For example, researchers coated carbon nanotubes with polystyrene by first polymerizing polystyrene via chain radical polymerization and subsequently mixing it at 130 °C with carbon nanotubes to generate radicals and graft them onto the walls of carbon nanotubes. Chain growth polymerization ("grafting to") synthesizes a polymer with predetermined properties. Purification of the polymer can be used to obtain a more uniform length distribution before grafting. Conversely, "grafting from", with radical polymerization techniques such as atom transfer radical polymerization (ATRP) or nitroxide-mediated polymerization (NMP), allows rapid growth of high molecular weight polymers.

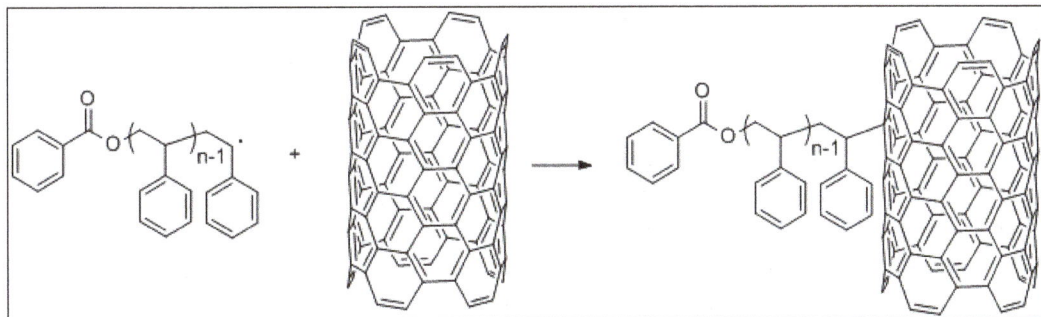

Grafting of a polystyrene free radical onto a single-walled carbon nanotube.

Radical polymerization also aids synthesis of nanocomposite hydrogels. These gels are made of water-swellable nano-scale clay (especially those classed as smectites) enveloped by a network polymer. They are often biocompatible and have mechanical properties (such as flexibility and strength) that promise applications such as synthetic tissue. Synthesis involves free radical polymerization. The general synthesis procedure is depicted in figure. Clay is dispersed in water, where it forms very small, porous plates. Next the initiator and a catalyst are added, followed by adding the organic monomer, generally an acrylamide or acrylamide derivative. The initiator is chosen to have stronger interaction with the clay than the organic monomer, so it preferentially adsorbs to the clay surface. The mixture and water solvent is heated to initiate polymerization. Polymers grow off the initiators that are in turn bound to the clay. Due to recombination and disproportionation reactions, growing polymer chains bind to one another, forming a strong, cross-linked network polymer, with clay particles acting as branching points for multiple polymer chain segments. Free radical polymerization used in this context allows the synthesis of polymers from a wide variety of substrates (the chemistries of suitable clays vary). Termination reactions unique to chain growth polymerization produce a material with flexibility, mechanical strength and biocompatibility.

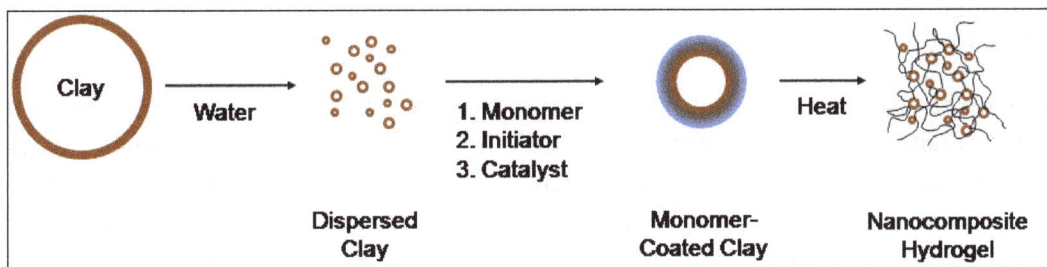

General synthesis procedure for a nanocomposite hydrogel.

Electronics

The radical polymer glass PTMA is about 10 times more electrically conductive than common semiconducting polymers. PTMA is in a class of electrically active polymers that could find use in transparent solar cells, antistatic and antiglare coatings for mobile phone displays, antistatic coverings for aircraft to protect against lightning strikes, flexible flash drives, and thermoelectric devices, which convert electricity into heat and the reverse. Widespread practical applications require increasing conductivity another 100 to 1,000 times.

The polymer was created using deprotection, which involves replacing a specific hydrogen atom in the pendant group with an oxygen atom. The resulting oxygen atom in PTMA has one unpaired electron in its outer shell, making it amenable to transporting charge. The deprotection step can lead to four distinct chemical functionalities, two of which are promising for increasing conductivity.

Living Polymerization

In polymer chemistry, living polymerization is a form of chain growth polymerization where the ability of a growing polymer chain to terminate has been removed. This can be accomplished in a variety of ways. Chain termination and chain transfer reactions are absent and the rate of chain initiation is also much larger than the rate of chain propagation. The result is that the polymer chains grow at a more constant rate than seen in traditional chain polymerization and their lengths remain very similar (i.e. they have a very low polydispersity index). Living polymerization is a popular method for synthesizing block copolymers since the polymer can be synthesized in stages, each stage containing a different monomer. Additional advantages are predetermined molar mass and control over end-groups.

Living polymerization is desirable because it offers precision and control in macromolecular synthesis. This is important since many of the novel/useful properties of polymers result from their microstructure and molecular weight. Since molecular weight and dispersity are less controlled in non-living polymerizations, this method is more desirable for materials design

In many cases, living polymerization reactions are confused or thought to be synonymous with controlled polymerizations. While these polymerization reactions are very similar, there is a distinct difference in the definitions of these two reactions. While living polymerizations are defined as polymerization reactions where termination or chain transfer is eliminated, controlled polymerization reactions are reactions where termination is suppressed, but not eliminated, through the introduction of a dormant state of the polymer. However, this distinction is still up for debate in the literature.

The main living polymerization techniques are:

- Living anionic polymerization.

- Living cationic polymerization.

- Living ring-opening metathesis polymerization.

- Living free radical polymerization.

- Living chain-growth polycondensations.

Fast Rate of Initiation: Low Polydispersity

One of the key characteristics of a living polymerization is that the chain termination and transfer reactions are essentially eliminated from the four elementary reactions of chain-growth polymerization leaving only initiation and (chain) propagation reactions.

A key characteristic of living polymerization is that the rate of initiation (meaning the dormant chemical species generates the active chain propagating species) is much faster than the rate of chain propagation. Thus all of the chains grow at the same rate (the rate of propagation).

The high rate of initiation (together with absence of termination) results in low (or narrow) polydispersity index (PDI), an indication of the broadness in the distribution of polymer chains (Living polymers) The extended lifetime of the propagating chain allowing for co-block polymer formation and end group functionalization to be performed on the living chain. These factors also allow predictable molecular weights, expressed as the number average molecular weight (M_n). For an ideal living system, assuming efficiency for generating active species is 100%, where each initiator generates only one active species the Kinetic chain length (average number of monomers the active species reacts with during its lifetime) at a given time can be estimated by knowing the concentration of monomer remaining. The number average molecular weight, M_n, increases linearly with percent conversion during a living polymerization:

$$v = \frac{[M]_0 - [M]}{[I]_0}$$

Techniques

Living Anionic Polymerization

As early as 1936, Karl Ziegler proposed that anionic polymerization of styrene and butadiene by consecutive addition of monomer to an alkyl lithium initiator occurred without chain transfer or termination. Twenty years later, living polymerization was demonstrated by Szwarc through the anionic polymerization of styrene in THF using sodium naphthalenide as an initiator.

The naphthalene anion initiate polymerization by reducing styrene to its radical anion, which dimerizes to the dilithiodiphenylbutane, which then initiates the polymerization.

These experiments relied on Szwarc's ability to control the levels of impurities which would destroy the highly reactive organometallic intermediates.

Living α-olefin Polymerization

α-olefins can be polymerized through an anionic coordination polymerization in which the metal center of the catalyst is considered the counter cation for the anionic end of the alkyl chain (through a M-R coordination). Ziegler-Natta initiators were developed in the mid-1950s and are heterogeneous initiators used in the polymerization of alpha-olefins. Not only were these initiators the first to achieve relatively high molecular weight poly(1-alkenes) (currently the most widely produced thermoplastic in the world PE(Polyethylene) and PP (Polypropylene) but the initiators were also capable of stereoselctive polymerizations which is attributed to the chiral Crystal structure of the heterogeneous initiator. Due to the importance of this discovery Ziegler and Natta were presented with the 1963 Nobel Prize in chemistry. Although the active species formed from the Ziegler-Natta initiator generally have long lifetimes (on the scale of hours or longer) the lifetimes of the propagating chains are shortened due to several chain transfer pathways (Beta-Hydride elimination and transfer to the co-initiator) and as a result are not considered living.

Metallocene initiators are considered as a type of Ziegler-Natta initiators due to the use of the two-component system consisting of a transition metal and a group I-III metal co-initiator (for example methylalumoxane (MAO) or other alkyl aluminum compounds). The metallocene initiators form homogeneous single site catalysts that were initially developed to study the impact that the catalyst structure had on the resulting polymers structure/properties; which was difficult for multi-site heterogeneous Ziegler-Natta initiators. Owing to the discrete single site on the metallocene catalyst researchers were able to tune and relate how the ancillary ligand (those not directly involved in the chemical transformations) structure and the symmetry about the chiral metal center affect the microstructure of the polymer. However, due to chain breaking reactions (mainly Beta-Hydride elimination) very few metallocene based polymerizations are known.

By tuning the steric bulk and electronic properties of the ancillary ligands and their substituents a class of initiators known as chelate initiators (or post-metallocene initiators) have been successfully used for stereospecific living polymerizations of alpha-olefins. The chelate initiators have a high potential for living polymerizations because the ancillary ligands can be designed to discourage or inhibit chain termination pathways. Chelate initiators can be further broken down based on the ancillary ligands; ansa-cyclopentyadienyl-amido initiators, alpha-diimine chelates and phenoxy-imine chelates.

Ansa-cyclopentadienyl-amido (CpA) Initiators

CpA initiators have one cyclopentadienyl substituent and one or more nitrogen substituents coordinated to the metal center (generally a Zr or Ti) (Odian). The dimethyl(pentamethylcyclopentyl)zirconium acetamidinate in figure has been used for a stereospecific living polymerization of 1-hexene at −10 deg C. The resulting poly(1-hexene) was isotactic (stereohemistry is the same between adjacent repeat units) confirmed by ^{13}C-NMR. The multiple trials demonstrated a controllable and predictable (from catalyst to monomer ratio) M_n with low Đ. The polymerization was further confirmed to be living by sequentially adding 2 portions of the monomer, the second portion was added after the first portion was already polymerized, and monitoring the Đ and M_n of the chain. The resulting polymer chains complied with the predicted M_n (with the total monomer concentration = portion 1 +2) and showed low Đ suggesting the chains were still active, or living, as the second portion of monomer was added.

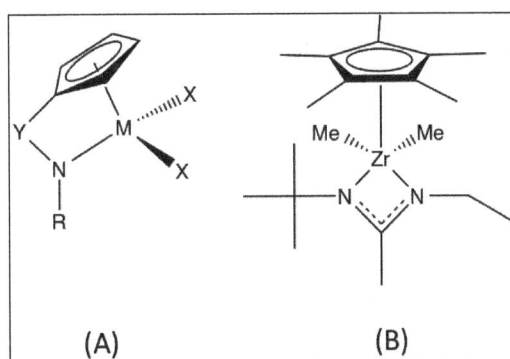

(A)   (B)

a) Shows the general form of CpA initiators with one Cp ring and a coordinated Nitrogen
b) Shows the CpA initiator used in the living polymerization of 1-hexene.

α-diimine Chelate Initiators

α-diimine chelate initiators are characterized by having a diimine chelating ancillary ligand structure and which is generally coordinated to a late transition (i.e. Ni and Pd) metal center.

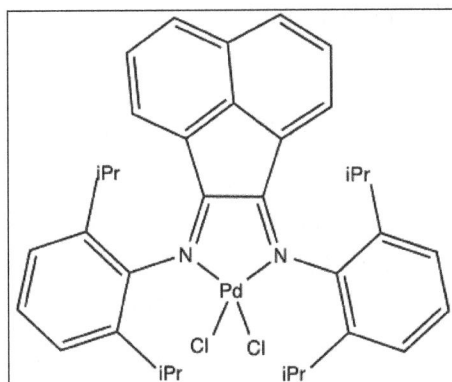

Brookhart et al. did extensive work with this class of catalysts and reported living polymerization for α-olefins and demonstrated living α-olefin carbon monoxide alternating copolymers.

Living Cationic Polymerization

Monomers for living cationic polymerization are electron-rich alkenes such as vinyl ethers, isobutylene, styrene, and N-vinylcarbazole. The initiators are binary systems consisting of an electrophile and a Lewis acid. The method was developed around 1980 with contributions from Higashimura, Sawamoto and Kennedy. Typically, generating a stable carbocation for a prolonged period of time is difficult, due to the possibility for the cation to be quenched by a β-protons attached to another monomer in the backbone, or in a free monomer. Therefore, a different approach is taken.

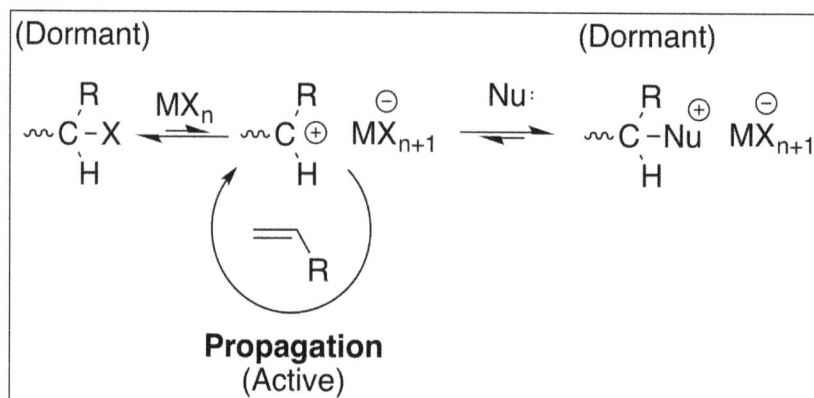

In this example, the carbocation is generated by the addition of a Lewis acid co-initiator, along with the halogen "X" already on the polymer, which ultimately generates the carbocation in a weak equilibrium. This equilibrium heavily favors the dormant state, thus leaving little time for permanent quenching or termination by other pathways. In addition, a weak nucleophile (Nu:) can also added to reduce the concentration of active species even further, thus keeping the polymer "living". However, it is important to note that by definition, the polymers described in this example are not technically living due to the introduction of a dormant state, as termination has only been decreased, not eliminated. But, they do operate similarly, and are used in similar applications to those of true living polymerizations.

Living Ring-opening Metathesis Polymerization

Given the right reaction conditions ring-opening metathesis polymerization (ROMP) can be rendered living. The first such systems were described by Robert H. Grubbs in 1986 based on norbornene and Tebbe's reagent and in 1978 Grubbs together with Richard R. Schrock describing living polymerization with a tungsten carbene complex.

Generally, ROMP reactions involve the conversion of a cyclic olefin with significant ring-strain (>5 kcal/mol), such as cyclobutene, norbornene, cyclopentene, etc., to a polymer that also contains double bonds. The important thing to note about ring-opening metathesis polymerizations is that the double bond is usually maintained in the backbone, which can allow it to be considered "living" under the right conditions.

For a ROMP reaction to be considered "living", several guidelines must be met:

1. Fast and complete initiation of the monomer. This means that the rate at which an initiating agent activates the monomer for polymerization, must happen very quickly.

2. How many monomers make up each polymer (the degree of polymerization) must be related linearly to the amount of monomer you started with.

3. The dispersity of the polymer must be < 1.5. In other words, the distribution of how long your polymer chains are in your reaction must be very low.

With these guidelines in mind, it allows you to create a polymer that is well controlled both in content (what monomer you use) and properties of the polymer (which can be largely attributed to polymer chain length). It is important to note that living ring-opening polymerizations can be anionic *or* cationic.

Because living polymers have had their termination ability removed, this means that once your monomer has been consumed, the addition of more monomer will result in the polymer chains continuing to grow until all of the additional monomer is consumed. This will continue until the metal catalyst at the end of the chain is intentionally removed by the addition of a quenching agent. As a result, it may potentially allow one to create a block or gradient copolymer fairly easily and accurately. This can lead to a high ability to tune the properties of the polymer to a desired application (electrical/ionic conduction, etc).

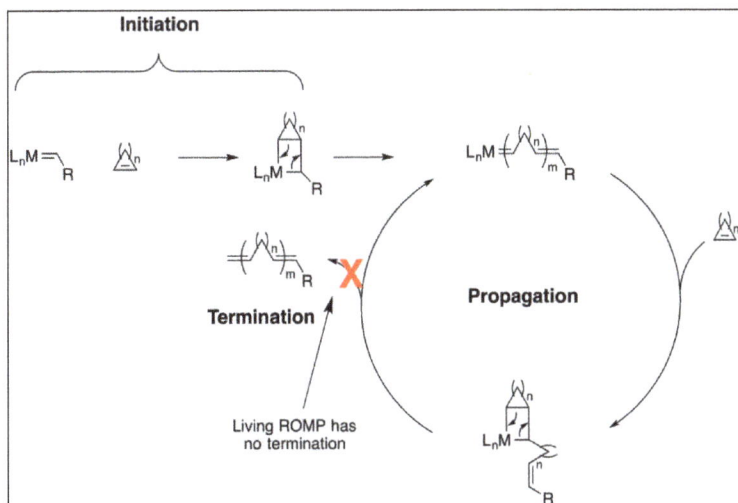

"Living" Free Radical Polymerization

Starting in the 1970s several new methods were discovered which allowed the development of living polymerization using free radical chemistry. These techniques involved catalytic chain transfer polymerization, iniferter mediated polymerization, stable free radical mediated polymerization (SFRP), atom transfer radical polymerization (ATRP), reversible addition-fragmentation chain transfer (RAFT) polymerization, and iodine-transfer polymerization.

In "living" radical polymerization (or controlled radical polymerization (CRP)) the chain breaking pathways are severely depressed when compared to conventional radical polymerization (RP) and CRP can display characteristics of a living polymerization. However, since chain termination is not absent, but only minimized, CRP technically does not meet the requirements imposed by IUPAC for a living polymerization. This issue has been up for debate the view points of different researchers can be found in a special issue of the Journal of Polymer Science titled Living or Controlled? The issue has not yet been resolved in the literature so it is often denoted as a "living" polymerization, quasi-living polymerization, pseudo-living and other terms to denote this issue.

There are two general strategies employed in CRP to suppress chain breaking reactions and promote fast initiation relative to propagation. Both strategies are based on developing a dynamic equilibrium amongst an active propagating radical and a dormant species.

The first strategy involves a reversible trapping mechanism in which the propagating radical undergoes an activation/deactivation (i.e. Atom-transfer radical-polymerization) process with a species X. The species X is a persistent radical, or a species that can generate a stable radical, that cannot terminate with itself or propagate but can only reversibly "terminate" with the propagating radical (from the propagating polymer chain)P*. P* is a radical species that can propagate (k_p) and irreversibly terminate (k_t) with another P*. X is normally a nitroxide (i.e. TEMPO used in Nitroxide Mediated Radical Polymerization) or an organometallic species. The dormant species (P_n-X) can be activated to regenerate the active propagating species (P*) spontaneously, thermally, using a catalyst and optically.

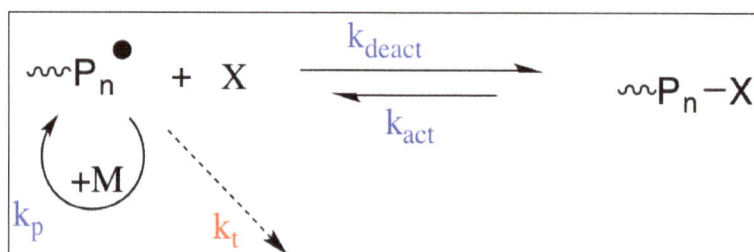

The second strategy is based on a degenerative transfer (DT) of the propagating radical between transfer agent that acts as a dormant species (i.e. Reversible addition–fragmentation chain-transfer polymerization). The DT based CRP's follow the conventional kinetics of radical polymerization, that is slow initiation and fast termination, but the transfer agent (Pm-X or Pn-X) is present in a much higher concentration compared to the radical initiator. The propagating radical species undergoes a thermally neutral exchange with the dormant transfer agent through atom transfer, group transfer or addition fragment chemistry.

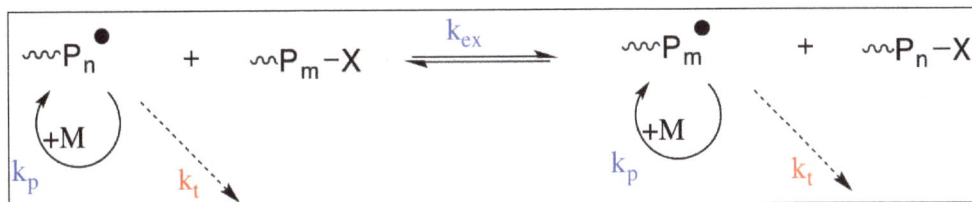

Living Chain-growth Polycondensations

Chain growth polycondensation polymerizations were initially developed under the premise that a change in substituent effects of the polymer, relative to the monomer, causes the polymers end group to be more reactive this has been referred to as "reactive intermediate polycondensation". The essential result is monomers preferentially react with the activated polymer end groups over reactions with other monomers. This preferred reactivity is the fundamental difference when categorizing a polymerization mechanism as chain-growth as opposed to step-growth in which the monomer and polymer chain end group have equal reactivity (the reactivity is uncontrolled). Several strategies were employed to minimize monomer-monomer reactions (or self-condensation) and polymerizations with low Đ and controllable Mn have been attained by this mechanism for small molecular weight polymers. However, for high molecular weight polymer chains (i.e. small initiator to monomer ratio) the Mn is not easily to controlled, for some monomers, since self-condensation between monomers occurred more frequently due to the low propagating species concentration.

Catalyst-transfer Polycondensation

Catalyst transfer polycondensation (CTP) is a chain-growth polycondensation mechanism in which the monomers do not directly react with one another and instead the monomer will only react with the polymer end group through a catalyst-mediated mechanism. The general process consists of the catalyst activating the polymer end group followed by a reaction of the end group with a 2nd incoming monomer. The catalyst is then transferred to the elongated chain while activating the end group.

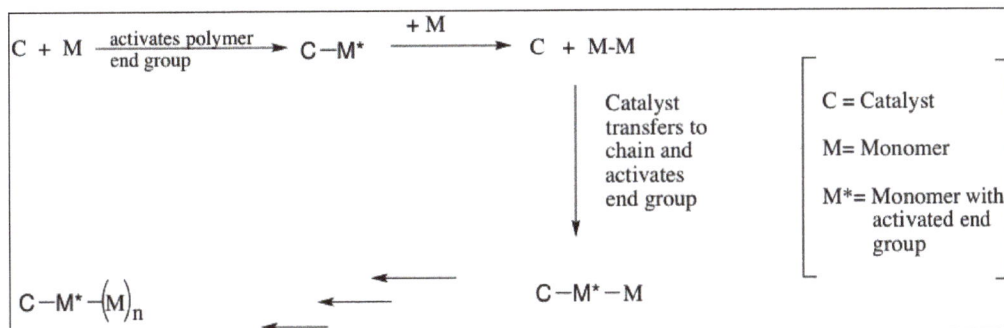

Catalyst transfer polycondensation allows for the living polymerization of π-conjugated polymers and was discovered by Tsutomu Yokozawa in 2004 and Richard McCullough.

In CTP the propagation step is based on organic cross coupling reactions (i.e. Kumada coupling, Sonogashira coupling, Negishi coupling) top form carbon carbon bonds between difunctional monomers. When Yokozawa and McCullough independently discovered the polymerization using a metal catalyst to couple a Grignard reagent with an organohalide making a new carbon-carbon bond. The mechanism below shows the formation of poly(3-alkylthiophene) using a Ni initiator (L_n can be 1,3-Bis(diphenylphosphino)propane (dppp)) and is similar to the conventional mechanism for Kumada coupling involving an oxidative addition, a transmetalation and a reductive elimination step. However, there is a key difference, following reductive elimination in CTP, an associative complex is formed (which has been supported by intra-/intermolecular oxidative addition competition experiments) and the subsequent oxidative addition occurs between the metal center and the associated chain (an intramolecular pathway). Whereas in a coupling reaction the newly formed alkyl/aryl compound diffuses away and the subsequent oxidative addition occurs between an incoming Ar-Br bond and the metal center. The associative complex is essential to for polymerization to occur in a living fashion since it allows the metal to undergo a preferred intramolecular oxidative addition and remain with a single propagating chain (consistent with chain-growth mechanism), as opposed to an intermolecular oxidative addition with other monomers present in the solution (consistent with a step-growth, non-living, mechanism). The monomer scope of CTP has been increasing since its discovery and has included poly(phenylene)s, poly(fluorine)s, poly(selenophene)s and poly(pyrrole)s.

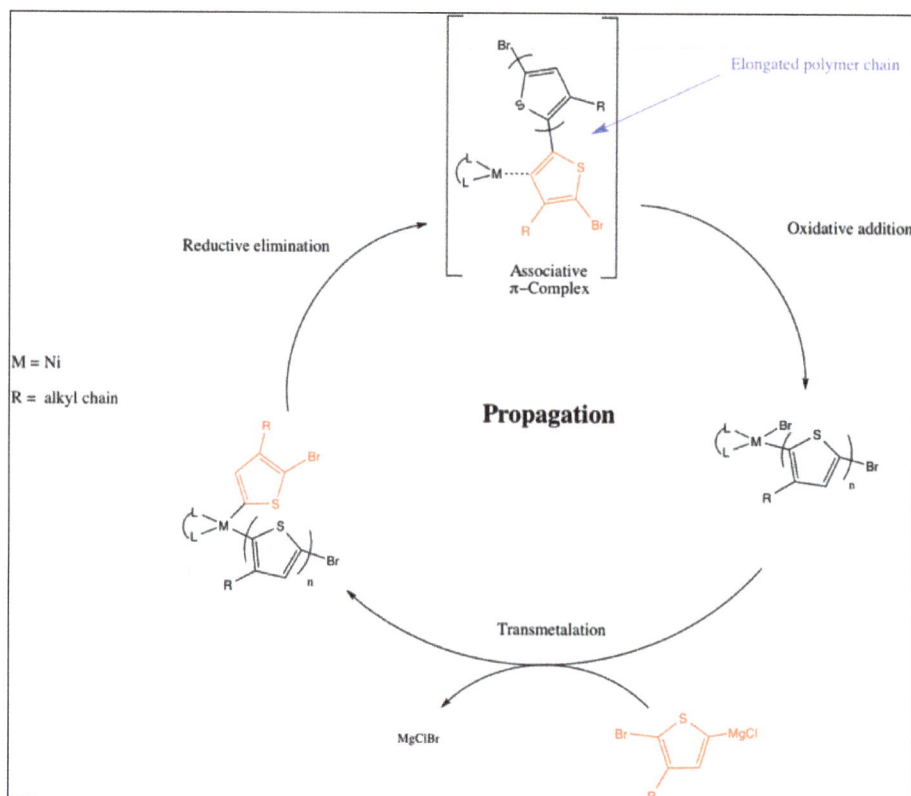

Living Group-transfer Polymerization

Group-transfer polymerization also has characteristics of living polymerization. It is applied to alkylated methacrylate monomers and the initiator is a silyl ketene acetal. New monomer adds to the initiator and to the active growing chain in a Michael reaction. With each addition of a monomer group the trimethylsilyl group is transferred to the end of the chain. The active chain-end is not ionic as in anionic or cationic polymeriation but is covalent. The reaction can be catalysed by bifluorides and bioxyanions such as tris(dialkylamino)sulfonium bifluoride or tetrabutyl ammonium bibenzoate. The method was discovered in 1983 by O.W. Webster and the name first suggested by Barry Trost.

Applications

Living polymerizations are used in the commercial synthesis of many polymers.

Copolymer Synthesis and Applications

Copolymers are polymers consisting of multiple different monomer species, and can be arranged in various orders, three of which are seen in the figure below.

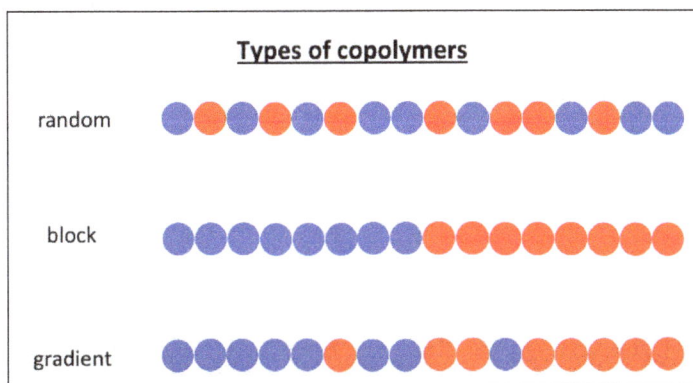

While there exist others (alternating copolymers, graft copolymers, and stereoblock copolymers), these three are more common in the scientific literature. In addition, block copolymers can exist as many types, including triblock (A-B-A), alternating block (A-B-A-B-A-B), etc.

Of these three types, block and gradient copolymers are commonly synthesized through living polymerizations, due to the ease of control living polymerization provides. Copolymers are highly desired due to the increased flexibility of properties a polymer can have compared to their homopolymer counterparts. The synthetic techniques used range from ROMP to generic anionic or cationic living polymerizations.

Copolymers, due to their unique tunability of properties, can have a wide range of applications. One example (of many) is nano-scale lithography using block copolymers. One used frequently is a block copolymer made of polystyrene and poly(methyl methacrylate) (abbreviated PS-*b*-PMMA). This copolymer, upon proper thermal and processing conditions, can form cylinders on the order of a few tens of nanometers in diameter of PMMA, surrounded by a PS matrix. These cylinders can then be etched away under high exposure to UV light and acetic acid, leaving a porous PS matrix.

The unique property of this material is that the size of the pores (or the size of the PMMA cylinders) can be easily tuned by the ratio of PS to PMMA in the synthesis of the copolymer. This can be easily tuned due to the easy control given by living polymerization reactions, thus making this technique highly desired for various nanoscale patterning of different materials for applications to catalysis, electronics, etc.

Living Cationic Polymerization

Living cationic polymerization is a living polymerization technique involving cationic propagating species. It enables the synthesis of very well defined polymers (low molar mass distribution) and of polymers with unusual architecture such as star polymers and block copolymers and living cationic polymerization is therefore as such of commercial and academic interest.

In carbocationic polymerization the active site is a carbocation with a counterion in close proximity. The basic reaction steps are:

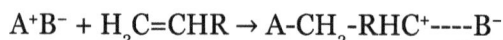

$$A^+B^- + H_2C=CHR \rightarrow A\text{-}CH_2\text{-}RHC^+\text{----}B^-$$

- Chain propagation:

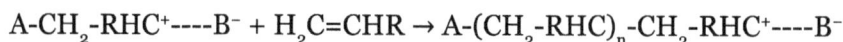

$$A\text{-}CH_2\text{-}RHC^+\text{----}B^- + H_2C=CHR \rightarrow A\text{-}(CH_2\text{-}RHC)_n\text{-}CH_2\text{-}RHC^+\text{----}B^-$$

- Chain termination:

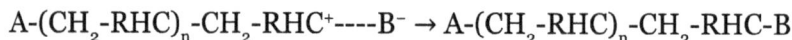

$$A\text{-}(CH_2\text{-}RHC)_n\text{-}CH_2\text{-}RHC^+\text{----}B^- \rightarrow A\text{-}(CH_2\text{-}RHC)_n\text{-}CH_2\text{-}RHC\text{-}B$$

- Chain transfer:

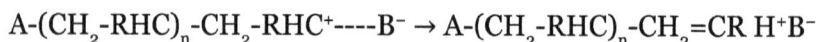

$$A\text{-}(CH_2\text{-}RHC)_n\text{-}CH_2\text{-}RHC^+\text{----}B^- \rightarrow A\text{-}(CH_2\text{-}RHC)_n\text{-}CH_2=CR\ H^+B^-$$

Living cationic polymerization is characterised by defined and controlled initiation and propagation while minimizing side-reactions termination and chain transfer. Transfer and termination do occur but in ideal living systems the active ionic propagating species are in chemical equilibrium with the dormant covalent species with an exchange rate much faster than the propagation rate. Solution methods require rigorous purification of monomer and solvent although conditions are not as strict as in anionic polymerization.

Common monomers are vinyl ethers, alpha-methyl vinyl ethers, isobutene, styrene, methylstyrene and N-vinylcarbazole. The monomer is nucleophilic and substituents should be able to stabilize a positive carbocationic charge. For example, para-methoxystyrene is more reactive than styrene itself.

Initiation takes place by an initiation/coinitiation binary system, for example an alcohol and a Lewis acid. The active electrophile is then a proton and the counter ion the remaining alkoxide which is stabilized by the Lewis acid. With organic acetates such as cumyl acetate the initiating species is the carbocation R^+ and the counterion is the acetate anion. In the iodine/HI system the electrophile is again a proton and the carbocation is stabilized by the triiodide ion. Polymerizations with Diethylaluminium chloride rely on trace amounts of water. A proton is then accompanied by the counterion Et_2Al-$ClOH^-$. With tert-butyl chloride Et_2AlCl abstracts a chlorine atom to form the tert-butyl carbocation as the electrophile. Efficient initiators that resemble the monomer are called cationogens. Termination and chain transfer are minimized when the initiator counterion is both non-nucleophilic and non-basic. More polar solvents promote ion dissociation and hence increase molar mass.

Common additives are electron donors, salts and proton traps. Electron donors (e.g. nucleophiles, Lewis bases) for example dimethylsulfide and dimethylsulfoxide are believed to stabilize the carbocation. The addition of salt for example a tetraalkylammonium salt, prevents dissociation of the ion pair that is the propagating reactive site. Ion dissociation into free ions lead to non-living polymerization. Proton traps scavenge protons originating from protic impurities.

Isobutylene Polymerization

Living isobutylene polymerization typically takes place in a mixed solvent system comprising a non-polar solvent, such as hexane, and a polar solvent, such as chloroform or dichloromethane, at temperatures below 0 °C. With more polar solvents polyisobutylene solubility becomes a problem. Initiators can be alcohols, halides and ethers. Co-initiators are boron trichloride, tin tetrachloride and organoaluminum halides. With ethers and alcohols the true initiator is the chlorinated product. Polymer with molar mass of 160,000 g/mole and polydispersity index 1.02 can be obtained.

Vinyl Ether Polymerization

Vinyl ethers (CH_2=CHOR, R = methyl, ethyl, isobutyl, benzyl) are very reactive vinyl monomers. Studied systems are based on I_2/HI and on zinc halides zinc chloride, zinc bromide and zinc iodide.

Living Cationic Ring-opening Polymerization

In Living cationic ring-opening polymerization the monomer is a heterocycle such as an epoxide, THF, an oxazoline or an aziridine such as t-butylaziridine. The propagating species is not a carbocation but an oxonium ion. Living polymerization is more difficult to achieve because of the ease of termination by nucleophilic attack of a heteroatom in the growing polymer chain. Intramolecular termination is called backbiting and results in the formation of cyclic oligomers. Initiators are strong electrophiles such as triflic acid. Triflic anhydride is an initiator for bifunctional polymer.

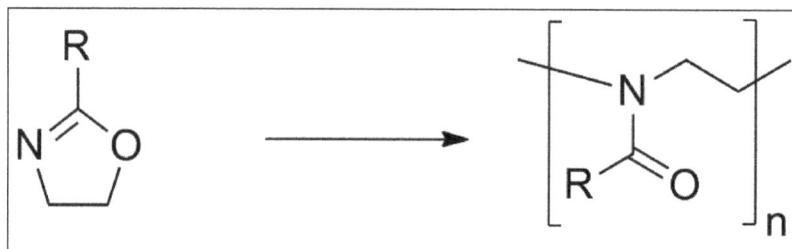

Living cationic ring-opening polymerization of 2oxazoline to poly(2oxazoline).

Living Free-radical Polymerization

Living free radical polymerization is a type of living polymerization where the active polymer chain end is a free radical. Several methods exist. IUPAC recommends to use the term "reversible-deactivation radical polymerization" instead of "living free radical polymerization", though the two terms are not synonymous.

Reversible-deactivation Polymerization

There is a mode of polymerization referred to as reversible-deactivation polymerization which is distinct from living polymerization, despite some common features. Living polymerization requires a complete absence of termination reactions, whereas reversible-deactivation polymerization may contain a similar fraction of termination as conventional polymerization with the same concentration of active species. Some important aspects of these are compared in the table.

Comparison of radical polymerization processes			
Property	Standard radical polymerization	Living polymerization	Reversible-deactivation polymerization
Concn. of initiating species	Falls off only slowly	Falls off very rapidly	Falls off very rapidly
Concn. of chain carriers (Number of growing chains)	Instantaneous steady state (Bodenstein approximation applies) decreasing throughout reaction	Constant throughout reaction	Constant throughout reaction
Lifetime of growing chains	$\sim 10^{-3}$ s	Same as reaction duration	Same as reaction duration
Main form of termination	Radical combination or radical disproportionation	Termination reactions are precluded	Termination reactions are not precluded

Degree of polymerization	Broad range ($Đ \geq 1.5$) Schulz-Zimm distribution	Narrow range ($Đ < 1.5$) Poisson distribution	Narrow range ($Đ < 1.5$) Poisson distribution
Dormant states	None	Rare	Predominant

Catalytic Chain Transfer and Cobalt Mediated Radical Polymerization

Catalytic chain transfer polymerization is not a strictly living form of polymerization. Yet it figures significantly in the development of later forms of living free radical polymerization. Discovered in the late 1970s in the USSR it was found that cobalt porphyrins were able to reduce the molecular weight during polymerization of methacrylates. Later investigations showed that the cobalt glyoxime complexes were as effective as the porphyrin catalysts and also less oxygen sensitive. Due to their lower oxygen sensitivity these catalysts have been investigated much more thoroughly than the porphyrin catalysts.

The major products of catalytic chain transfer polymerization are vinyl-terminated polymer chains. One of the major drawbacks of the process is that catalytic chain transfer polymerization does not produce macromonomers but instead produces addition fragmentation agents. When a growing polymer chain reacts with the addition fragmentation agent the radical end-group attacks the vinyl bond and forms a bond. However, the resulting product is so hindered that the species undergoes fragmentation, leading eventually to telechelic species.

These addition fragmentation chain transfer agents do form graft copolymers with styrenic and acrylate species however they do so by first forming block copolymers and then incorporating these block copolymers into the main polymer backbone.

While high yields of macromonomers are possible with methacrylate monomers, low yields are obtained when using catalytic chain transfer agents during the polymerization of acrylate and stryenic monomers. This has been seen to be due to the interaction of the radical centre with the catalyst during these polymerization reactions.

The reversible reaction of the cobalt macrocycle with the growing radical is known as cobalt carbon bonding and in some cases leads to living polymerization reactions.

Iniferter Polymerization

An iniferter is a chemical compound that simultaneously acts as initiator, transfer agent, and terminator (hence the name ini-fer-ter) in controlled free radical iniferter polymerizations, the most common is the dithiocarbamate type.

Stable Free Radical Mediated Polymerization

The two options of SFRP are nitroxide mediated polymerization (NMP) and verdazyl mediated polymerization (VMP), SFRP was discovered while using a radical scavenger called TEMPO when investigating the rate of initiation during free radical polymerization. When the coupling of the stable free radical with the polymeric radical is sufficiently reversible, termination is reversible, and the propagating radical concentration can be limited to levels that allow controlled polymerization. Similar to atom transfer radical polymerization, the equilibrium between dormant chains (those reversibly terminated with the stable free radical) and active chains (those with a radical capable of adding to monomer) is designed to heavily favor the dormant state. Further stable free radicals have also been explored for this polymerization reaction with lower efficiency.

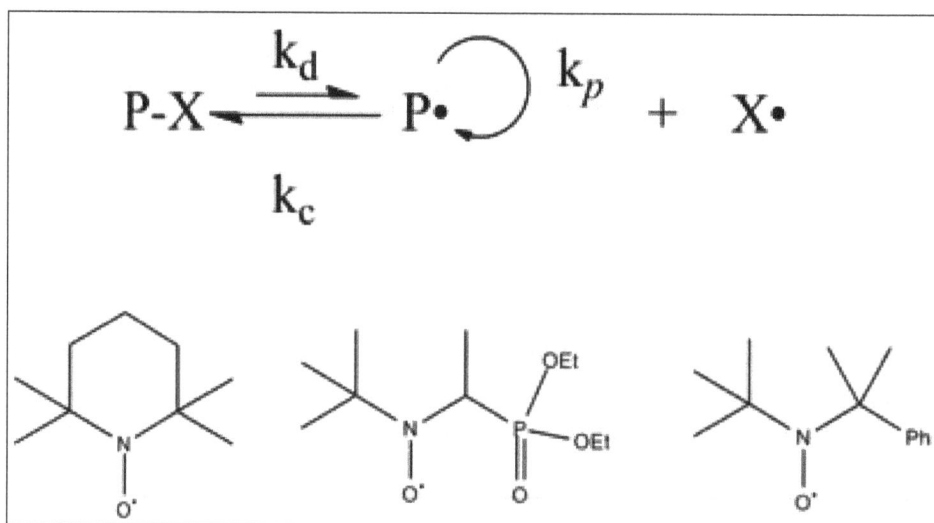

Atom Transfer Radical Polymerization (ATRP)

A review written by Matyjaszewski covers the developments in ATRP from 1995 to 2000. ATRP involves the chain initiation of free radical polymerization by a halogenated organic species in the presence of a metal halide. The metal has a number of different oxidation states that allows it to abstract a halide from the organohalide, creating a radical that then starts free radical polymerization. After inititation and propagation, the radical on the active chain terminus is reversibly terminated (with the halide) by reacting with the catalyst in its higher oxidation state. Thus, the redox process gives rise to an equilibrium between dormant (polymer-halide) and active (polymer-radical) chains. The equilibrium is designed to heavily favor the dormant state, which effectively reduces the radical concentration to a sufficiently low level to limit bimolecular coupling.

Obstacles associated with this type of reaction is the generally low solubility of the metal halide species, which results in limited availability of the catalyst. This is improved by the addition of a ligand, which significantly improves the solubility of the metal halide and thus the availability of the catalyst but complicates subsequent catalyst removal from the polymer product.

Reversible Addition Fragmentation Chain Transfer (RAFT) Polymerization

RAFT technology offers the benefit of being able to readily synthesize polymers with predetermined molecular weight and narrow molecular weight distributions over a wide range of monomers with reactive terminal groups that can be purposely manipulated, including further polymerization, with complex architecture.6 Furthermore, RAFT can be used in all modes of free radical polymerization: solution, emulsion and suspension polymerizations. Implementing the RAFT technique can be as simple as introducing a suitable chain transfer agent (CTA), known as a RAFT agent, into a conventional free radical polymerization reaction (must be devoid of oxygen, which terminates propagation). This CTA is the main species in RAFT polymerization. Generally it is a di- or tri-thiocarbonylthio compound (1), which produces the dormant form of the radical chains. Control in RAFT polymerization (scheme 1) is achieved in a far more complicated manner than the homolytic bond formation-bond cleavage of SFRP and ATRP. The CTA for RAFT polymerization must cautiously chosen because it has an effect on polymer length, chemical composition, rate of the reaction and the number of side reactions that may occur.

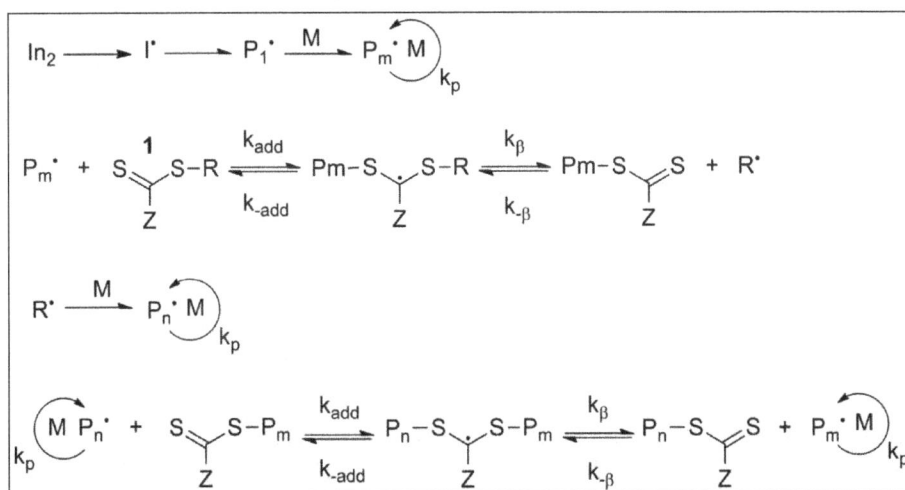

The mechanism of RAFT begins with a standard initiation step as homolytic bond cleavage of the initiator molecule yields a reactive free radical. This free radical then reacts with a molecule of the monomer to form the active center with additional molecules of monomer then adding in a sequential fashion to produce a growing polymer chain (Pn•). The propagating chain adds to the CTA to yield a radical intermediate. Fragmentation of this intermediate gives rise to either the original polymer chain (Pn•) or to a new radical (R•), which itself must be able to reinitiate polymerization. This free radical generates its own active center by reaction with the monomer and eventually a new propagating chain (Pm•) is formed. Ultimately, chain equilibration occurs in which there is a rapid equilibrium between the actively growing radicals and the dormant compounds, thereby allowing all of the chains to grow at the same rate. A limited amount of termination does occur; however, the effect of termination of polymerization kinetics is negligible.

The calculation of molecular weight for a synthesized polymer is relatively easy, in spite of the complex mechanism for RAFT polymerization. As stated before, during the equilibration step, all chains are growing at equal rates, or in other words, the molecular weight of the polymer increases linearly with conversion. Multiplying the ratio of monomer consumed to the concentration of the CTA used by the molecular weight of the monomer (mM) a reliable estimate of the number average molecular weight can be determined.

RAFT is a degenerative chain transfer process and is free radical in nature. RAFT agents contain di- or tri-thiocarbonyl groups, and it is the reaction with an initiator, usually AIBN, that creates a propagating chain or polymer radical. This polymer chain then adds to the C=S and leads to the formation of a stabilized radical intermediate. In an ideal system, these stabilized radical intermediates do not undergo termination reactions, but instead reintroduce a radical capable of reinitiation or propagation with monomer, while they themselves reform their C=S bond. The cycle of addition to the C=S bond, followed by fragmentation of a radical, continues until all monomer or initiator is consumed. Termination is limited in this system by the low concentration of active radicals and any termination that does occur is negligible. RAFT, invented by Rizzardo *et al.* at CSIRO and a mechanistically identical process termed Macromolecular Design via Interchange of Xanthates (MADIX), invented by Zard *et al.* at Rhodia were both first reported in 1998/early 1999.

Iodine-transfer Polymerization (ITP)

Iodine-transfer polymerization (ITP, also called ITRP), developed by Tatemoto and coworkers in the 1970s gives relatively low polydispersities for fluoroolefin polymers. While it has received relatively little academic attention, this chemistry has served as the basis for several industrial patents and products and may be the most commercially successful form of living free radical polymerization. It has primarily been used to incorporate iodine cure sites into fluoroelastomers.

The mechanism of ITP involves thermal decomposition of the radical initiator (AIBN), generating the initiating radical In•. This radical adds to the monomer M to form the species P_1•, which can propagate to P_m•. By exchange of iodine from the transfer agent R-I to the propagating radical P_m• a new radical R• is formed and P_m• becomes dormant. This species can propagate with monomer M to P_n•. During the polymerization exchange between the different polymer chains and the transfer agent occurs, which is typical for a degenerative transfer process.

Typically, iodine transfer polymerization uses a mono- or diiodo-perfluoroalkane as the initial chain transfer agent. This fluoroalkane may be partially substituted with hydrogen or chlorine. The energy of the iodine-perfluoroalkane bond is low and, in contrast to iodo-hydrocarbon bonds, its polarization small. Therefore, the iodine is easily abstracted in the presence of free radicals. Upon encountering an iodoperfluoroalkane, a growing poly(fluoroolefin) chain will abstract the iodine and terminate, leaving the now-created perfluoroalkyl radical to add further monomer. But the iodine-terminated poly(fluoroolefin) itself acts as a chain transfer agent. As in RAFT processes, as long as the rate of initiation is kept low, the net result is the formation of a monodisperse molecular weight distribution.

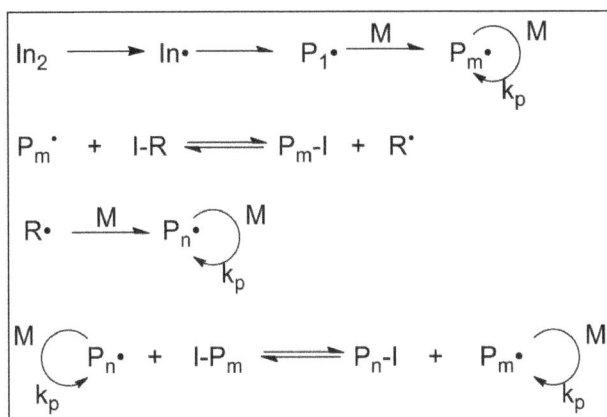

Use of conventional hydrocarbon monomers with iodoperfluoroalkane chain transfer agents has been described. The resulting molecular weight distributions have not been narrow since the energetics of an iodine-hydrocarbon bond are considerably different from that of an iodine-fluorocarbon bond and abstraction of the iodine from the terminated polymer difficult. The use of hydrocarbon iodides has also been described, but again the resulting molecular weight distributions were not narrow.

Preparation of block copolymers by iodine-transfer polymerization was also described by Tatemoto and coworkers in the 1970s.

Although use of living free radical processes in emulsion polymerization has been characterized as difficult, all examples of iodine-transfer polymerization have involved emulsion polymerization. Extremely high molecular weights have been claimed.

Listed below are some other less described but to some extent increasingly important living radical polymerization techniques.

Selenium-centered Radical-mediated Polymerization

Diphenyl diselenide and several benzylic selenides have been explored by Kwon *et al.* as photoiniferters in polymerization of styrene and methyl methacrylate. Their mechanism of control over polymerization is proposed to be similar to the dithiuram disulfide iniferters. However, their low transfer constants allow them to be used for block copolymer synthesis but give limited control over the molecular weight distribution.

Telluride-mediated Polymerization (TERP)

Telluride-mediated polymerization or TERP first appeared to mainly operate under a reversible chain transfer mechanism by homolytic substitution under thermal initiation. However, in a kinetic study it was found that TERP predominantly proceeds by degenerative transfer rather than 'dissociation combination'.

Alkyl tellurides of the structure Z-X-R, were Z=methyl and R= a good free radical leaving group, give the better control for a wide range of monomers, phenyl tellurides (Z=phenyl) giving poor control. Polymerization of methyl methacrylates are only controlled by ditellurides. The importance of X to chain transfer increases in the series O<S<Se<Te, makes alkyl tellurides effective in mediating control under thermally initiated conditions and the alkyl selenides and sulfides effective only under photoinitiated polymerization.

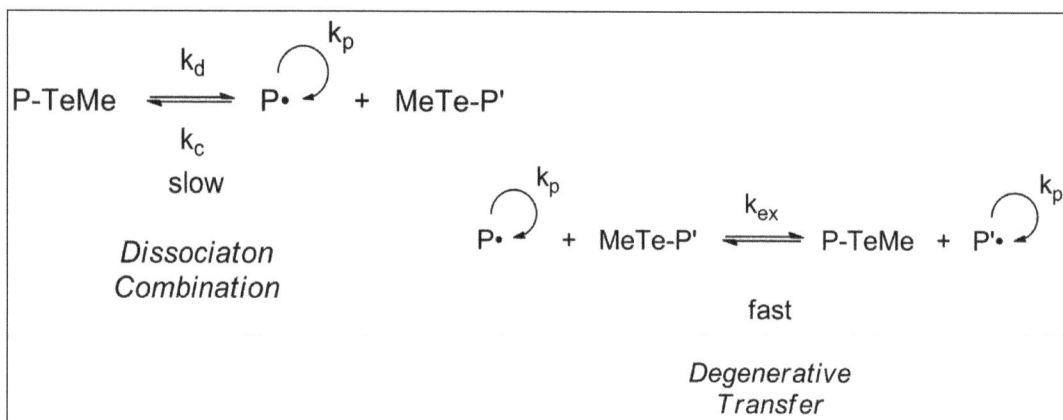

Stibine-mediated Polymerization

More recently Yamago *et al.* reported stibine-mediated polymerization, using an organostibine transfer agent with the general structure Z(Z')-Sb-R (where Z= activating group and R= free radical leaving group). A wide range of monomers (styrenics, (meth) acrylics and vinylics) can be controlled, giving narrow molecular weight distributions and predictable molecular weights under thermally initiated conditions. Yamago has also published a patent indicating that bismuth alkyls can also control radical polymerizations via a similar mechanism.

Coordination Polymerization

Coordination polymerisation is a form of polymerization that is catalyzed by transition metal salts and complexes.

Types of Coordination Polymerization of Alkenes

Heterogeneous Ziegler–Natta Polymerization

Coordination polymerization started in the 1950s with heterogeneous Ziegler–Natta catalysts based on titanium tetrachloride and organoaluminium co-catalyst. The mixing of $TiCl_4$ with trialkylaluminium complexes produces Ti(III)-containing solids that catalyze the polymerization of ethylene and propylene. The nature of the catalytic center has been of intense interest but remains uncertain. Many additives and variations have been reported for the original recipes.

Homogeneous Ziegler–Natta Polymerization

In some applications heterogeneous Ziegler–Natta polymerization has been superseded by homogeneous catalysts such as the Kaminsky catalyst discovered in the 1970s. The 1990s brought forward a new range of post-metallocene catalysts. Typical monomers are nonpolar ethylene and propylene. The development of coordination polymerization that enables copolymerization with polar monomers is more recent. Examples of monomers that can be incorporated are methyl vinyl ketones methyl acrylate and acrylonitrile.

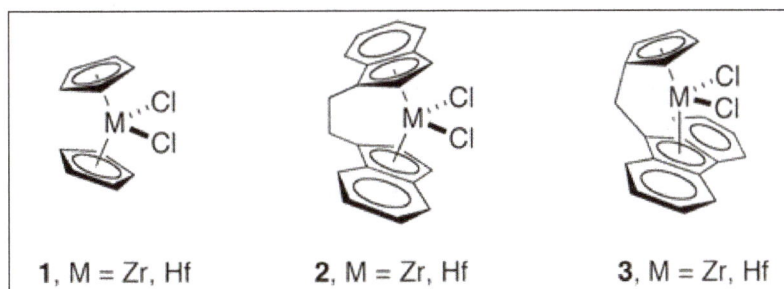

1, M = Zr, Hf   2, M = Zr, Hf   3, M = Zr, Hf

Illustrative metallocene-based coordination catalysts.

Kaminsky catalysts are based on metallocenes of group 4 metals (Ti, Zr, Hf) activated with methylaluminoxane (MAO). Polymerizations catalysed by metallocenes occur via the Cossee–Arlman mechanism. The active site is usually anionic but cationic coordination polymerization also exists.

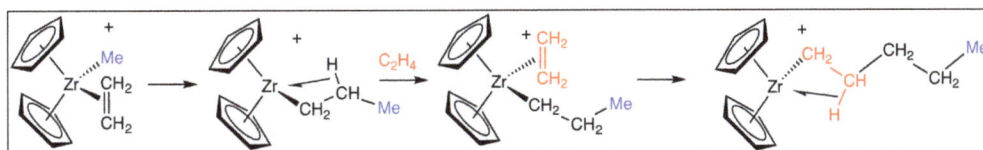

Simplified mechanism for Zr-catalyzed for ethylene polymerization.

Specialty Monomers

Many alkenes do not polymerize in the presence of Ziegler–Natta or Kaminsky catalysts. This problem applies to polar olefins such as vinyl chloride, vinyl ethers, and acrylate esters.

Butadiene Polymerization

The annual production of polybutadiene is 2.1 million tons (2000). The process employs a neodymium-based homogeneous catalyst.

Principles

Coordination polymerization has a great impact on the physical properties of vinyl polymers such as polyethylene and polypropylene compared to the same polymers prepared by other techniques such as free-radical polymerization. The polymers tend to be linear and not branched and have much higher molar mass. Coordination type polymers are also stereoregular and can be isotactic or syndiotactic instead of just atactic. This tacticity introduces crystallinity in otherwise amorphous polymers. From these differences in polymerization type the distinction originates between low-density polyethylene (LDPE), high-density polyethylene (HDPE) or even ultra-high-molecular-weight polyethylene (UHMWPE).

Coordination Polymerization of Other Substrates

Coordination polymerization can also be applied to non-alkene substrates. Dehydrogenative coupling of silanes of dihydro- and trihydrosilanes to polysilanes has been investigated, although the technology has not been commercialized. The process entails coordination and often oxidative addition of Si-H centers to metal complexes.

Lactides also polymerize in the presence of Lewis acidic catalysts to give polylactide:

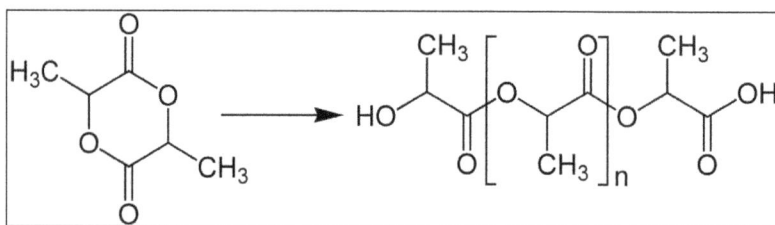

Chain-growth Polymerization

Chain polymerization is a chain reaction in which the growth of a polymer chain proceeds exclusively by reaction(s) between monomer(s) and active site(s) on the polymer chain with regeneration of the active site(s) at the end of each growth step.

It is a kind polymerization where an active center (free radical or ion) is formed, and a plurality of monomers can be polymerized together in a short period of time to form a macromolecule having a large molecular weight. In addition to the regenerated active sites of each monomer unit, polymer growth will only occur at one (or possibly more) endpoint.

Many common polymers can be obtained by chain polymerization such as Polyethylene (PE), polypropylene (PP), polyvinyl chloride (PVC), polymethyl methacrylate, polyacrylonitrile, polyvinyl acetate.

Typically, chain-growth polymerization can be understood with the chemical equation:

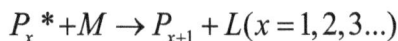

$$P_x * + M \rightarrow P_{x+1} + L (x = 1, 2, 3 ...)$$

In this equation, P is the polymer while x represents degree of polymerization, * means active center of chain-growth polymerization. M is the monomer which will reacted with active center. L is a low-molar-mass by-product obtained during chain propagation. Usually, for chain-growth polymerization, there is no by-product formed. However, there are still some exceptions. For example, amino acid N-carboxy anhydrides polymerizing to oxazolidine-2,5-diones.

Steps of Chain-growth Polymerization

Typically, chain polymerization must contain chain initiation and chain propagation. Chain transfer and chain termination do not always happened in a chain-growth polymerization.

Chain Initiation

Chain initiation is the process of initially generating a chain carrier (chain carriers are some intermediates such as radical and ions in chain propagation process) in a chain polymerization. According to different ways of energy dissipation, it can be divided into thermal initiation, high energy initiation, and chemical initiation, etc. Thermal initiation is a process that obtained energy and dissociated to homolytic cleavage to form active center by molecular thermal motion. High energy initiation refers to the generation of chain carriers by radiation. Chemical initiation is due to chemical initiator.

Chain Propagation

IUPAC defined chain propagation as an active center on the growing polymer molecule, which adds one monomer molecule to form a new polymer molecule which is one repeat unit longer with a new active center.

Chain Transfer

The polymerization process does not have to undergo chain transfer. Chain transfer means that in a chain polymerization, the active center of the polymer A takes an atom from B molecule and terminates. The B molecule produces a new active center instead. It can happen in free radical polymerization, ionic polymerization and coordination polymerization. Generally, chain transfer will generate by-product.

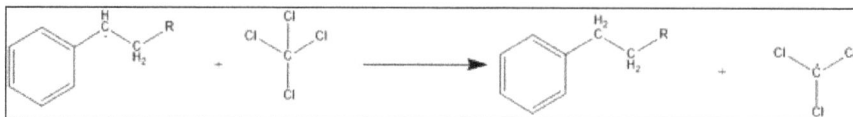

An example of chain transfer.

Chain Termination

Chain termination refers to in chain polymerization process, active center disappears, resulting in the termination of chain propagation. It is different from chain transfer. During the chain transfer process, the active point only shifts to another molecule but does not disappear.

Classes of Chain-growth Polymerization

Radical Polymerization

Based on definition from IUPAC, radical polymerization is a chain polymerization in which the kinetic-chain carriers are radicals. Usually, the growing chain end bears an unpaired electron. Free radicals can be initiated by many methods such as heating, redox reactions, ultraviolet radiation, high energy irradiation, electrolysis, sonication, and plasma. Free radical polymerization is very important in polymer chemistry. It is one of the most the most developed method in chain-growth polymerization. Currently, most polymers in our daily life are synthesized by free radical polymerization, such as polyethylene, polystyrene, polyvinyl chloride, polymethyl methacrylate, polyacrylonitrile, polyvinyl acetate, styrene butadiene rubber, nitrile rubber, neoprene, etc.

Ionic Polymerization

Based on IUPAC, ionic polymerization is a chain polymerization in which the kinetic-chain carriers are ions or ion pairs. It can be further divided into anionic polymerization and cationic polymerization. Ionic polymerization is widely used in our daily life. A lot of common polymers are generated by ionic polymerization such as butyl rubber, polyisobutylene, polyphenylene, polyoxymethylene, polysiloxane, polyethylene oxide, high density polyethylene, isotactic polypropylene, butadiene rubber, etc. Living anionic polymerization was developed since the 1950s, the chain will remain active indefinitely unless the reaction Is transferred or terminated deliberately, which realizes the control of molar weight and PDI.

Coordination Polymerization

Based on definition from IUPAC, coordination polymerization is a chain polymerization that involves the preliminary coordination of a monomer molecule with a chain carrier. The monomer is firstly coordinated with the transition metal active center, and then the activated monomer is inserted into the transition metal-carbon bond for chain growth. In some cases, coordination polymerization is also called insertion polymerization or complexing polymerization. Advanced coordination polymerizations can control the tacticity, molecular weight and PDI of the polymer effectively. In addition, the racemic mixture of the chiral metallocene can be separated into its enantiomers. The oligomerization reaction produces an optically active branched olefin using an optically active catalyst.

Living Polymerization

Living polymerization was first introduced by Szware in 1956. Based on definition from IUPAC, it is a chain polymerization from which chain transfer and chain termination are absent. As there is no chain-transfer and chain termination, the monomer in the system is consumed and the polymerization is stopped when the polymer chain remains active. Once the new monomer is added, the polymerization can proceed. Due to the low PDI and predictable molecular weight, living polymerization is at the forefront of polymer research. It could be further divided into living free radical polymerization, living ionic polymerization and living ring-opening metathesis polymerization, etc.

In the sheme, P is a polymer with degree of polymerization of n. X is the end atom or group of domaint chain. M is Mnomer. Y represents a catalyst, usually a metal complex. All of these three methods for living polymerization includes activation and deactivation process. The rates of activation and deactivation are written as k_{act} and k_{deact}. k_{act} and k_{deact} in the equilibrium could be further simplified as k_{exch}. K_p and k_t are propogation rate and termination rate. In living polymerizaiton system, the value of k_t is very small.

Ring-opening Polymerization

According to definition from IUPAC ring-opening polymerization is a polymerization in which a cyclic monomer yields a monomeric unit which is acyclic or contains fewer cycles than the monomer. Generally, the ring-opening polymerization is carried out under mild conditions, and the by-product is less than the polycondensation reaction and the high molecular weight polymer is easily obtained. Common ring-opening polymerization products includes polypropylene oxide, polytetrahydrofuran, polyepichlorohydrin, polyoxymethylene, polycaprolactam and polysiloxane.

Reversible-deactivation Polymerization

IUPAC stipulates that reversible-deactivation polymerization is a kind of chain polymerization, which is propagated by chain carriers that are deactivated reversibly, bringing them into active-dormant equilibria of which there might be more than one. An example of a reversible-deactivation polymerization is group-transfer polymerization.

Comparison to other Polymerization Methods

Previously, based on the difference between condensation reaction and addition reaction, Carothers classified polymerization as condensation polymerizations and addition polymerizations in 1929. However, Carothers' classification is not good enough in mechanism aspect, as in some case, addition polymerizations shows condensation features while condensation polymerization shows addition features. Then the classification was optimized as step-growth polymerization and chain-growth polymerization. Based on IUPAC recommendation, the names of step-growth polymerization and chain-growth polymerization were further simplified as polyaddition and chain polymerization.

Step-growth Polymerization

A step-growth reaction could happened between any two molecules with same or different degree of polymerization, usually monomers will form dimers, trimers, in matrix and finally react to long chain polymers. The mechanism of step-growth reaction is based on their functional group. Step-growth polymerization includes polycondensation and polyaddition. Polycondensation is a kind of polymerization whose chain growth is based on condensation reaction between two molecules with various degree of polymerization. The typical example are polyesters, polyamides and polyethers. It is sometimes confused by condensation previous definition of condensation polymerization. Polyaddition is a type of step-growth polymerization of which chain growth is based on addition reaction between two molecules of various degree of polymerization. The typical example for polyaddition is the synthesis of polyurethane. Compared to chain-growth polymerization, where the production of the growing chaingrowth is based on the reaction between polymer with active center and monomer, step-growth polymerization doesn't have initiator or termination. The monomer in step-growth polymerization will be consumed very quickly to dimer, trimer or oligomer. The degree of polymerization will increase steady during the whole polymerization process. On the other hand, in chain-growth polymerization, the monomer consumed steadily but the degree of polymerization could be increased very quickly after chain initiation. Compared to step-growth polymerization, living chain-growth polymerization shows low PDI, predictable molecular mass and controllable conformation. Some researchers are working on the transformation of two polymerization methods. Generally, polycondensation proceeds in a step-growth polymerization mode. Substituent effect, catalyst transfer and biphasic system could be used for inhibiting the activity of monomer, and further prevent monomers from reacting with each other. It could make polycondensation process go in a chain-growth polymerization mode.

Polycondensation

The chain growth of polycondensation is based on condensation reaction. Low-molarmass by-product will be formed during polymerization. It is a previous way to classify polymerization, which was introduced by Carothers in 1929. It is still used currently in some case. The step-growth polymerization with low-malar-mass by-product during chain growth is defined as polycondensation. The chain-growth polymerization a with low-malar-mass by-product during chain growth is recommended by IUPAC as "condensative chain polymerization".

Addition Polymerization

Addition polymerization is also a type of previous definition. The chain growth of addition polymerization is based addition reactions. There is no low-malar-mass by-product formed during polymerization. The step-growth polymerization based on addition reaction during chain growth is defined as polyaddition. Based on that definition, the addition polymerization contains both polyaddition and chain polymerization except condensative chain polymerization we used now.

Application

Chain polymerization is widely used in many aspects of life, including electronic devices, food packaging, catalyst carriers, medical materials, etc. At present, the world's highest yielding polymers such as polyethylene (PE), polyvinyl chloride (PVC), polypropylene (PP), etc. can be obtained by chain polymerization. In addition, some Carbon nanotube–polymer is used for electronical devices. Controlled living chain-growth conjugated polymerization will also enable the synthesis of well-defined advanced structures, including block copolymers. Their industrial applications extend to water purification, biomedical devices and sensors.

Anionic Addition Polymerization

Anionic addition polymerization is a form of chain-growth polymerization or addition polymerization that involves the polymerization of monomers initiated with anions. The type of reaction has many manifestations, but traditionally vinyl monomers are used. Often anionic polymerization involves living polymerizations, which allows control of structure and composition.

Monomer Characteristics

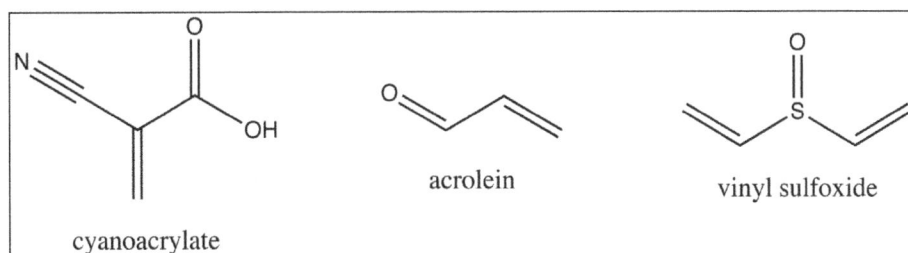

cyanoacrylate

acrolein

vinyl sulfoxide

Examples of polar monomers.

Two broad classes of monomers are susceptible to anionic polymerization. Vinyl monomers have the formula $CH_2=CHR$, the most important are styrene (R= C_6H_5), butadiene ($CH=CH_2$), and isoprene (R = $C(Me)=CH_2$). A second major class of monomers are acrylate esters, such as acrylonitrile, methacrylate, cyanoacrylate, and acrolein. Other vinyl monomers include vinylpyridine, vinyl sulfone, vinyl sulfoxide, vinyl silanes.

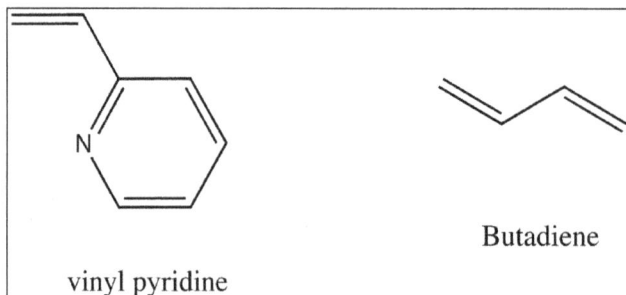

vinyl pyridine

Butadiene

Examples of vinyl monomers.

Cyclic Monomers

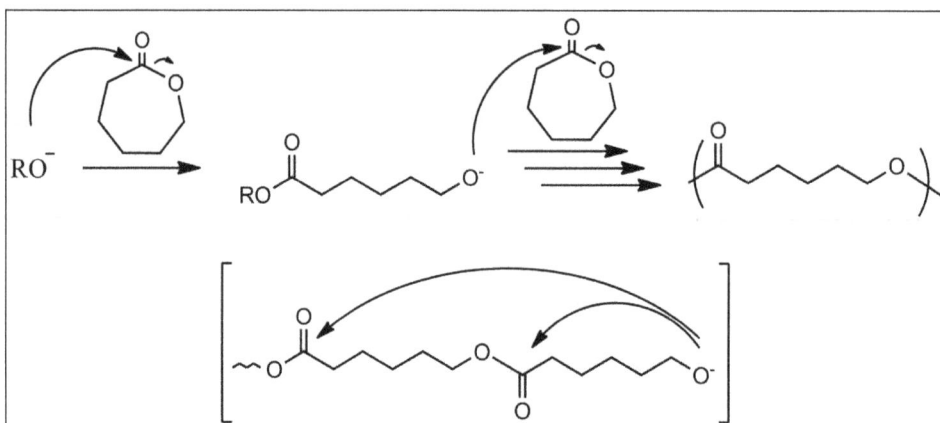

The anionic ring-opening polymerization of ε-caprolactone, initiated by alkoxide.

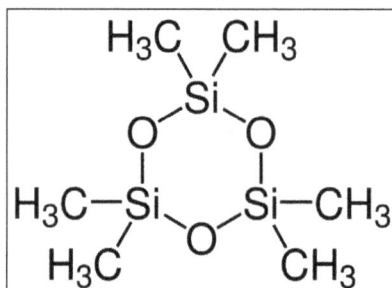

Hexamethylcyclotrisiloxane is a cyclic monomer that is susceptible to anionic polymerization to siloxane polymers.

Many cyclic compounds are susceptible to ring-opening polymerization. Epoxides, cyclic trisiloxanes, some lactones, lactides, cyclic carbonates, and amino acid N-carboxyanhydrides.

In order for polymerization to occur with vinyl monomers, the substituents on the double bond must be able to stabilize a negative charge. Stabilization occurs through delocalization of the negative charge. Because of the nature of the carbanion propagating center, substituents that react with bases or nucleophiles either must not be present or be protected.

Initiation

Initiators is selected based on the reactivity of the monomers. Highly electrophilic monomers such as cyanoacrylates require only weakly nucleophilic initiators, such as amines, phosphines, or even halides. Less reactive monomers such as styrene require powerful nucleophiles such as butyl lithium. Reaction of intermediate strength are used for monomers of intermediate reactivity such as vinylpyridine.

The solvent used in anionic addition polymerizations are determined by the reactivity of both the initiator and nature of the propagating chain end. Anionic species with low reactivity, such as heterocyclic monomers, can use a wide range of solvents.

Initiation by Electron Transfer

Initiation of styrene polymerization with sodium naphthalene proceeds by electron transfer from the naphthalene radical anion to the monomer. The resulting radical dimerizes to give a dilithio compound, which then functions as the initiator. Polar solvents are necessary for this type of initiation both for stability of the anion-radical and to solvate the cation species formed. The anion-radical can then transfer an electron to the monomer. Initiation can also involve the transfer of an electron from the alkali metal to the monomer to form an anion-radical. Initiation occurs on the surface of the metal, with the reversible transfer of an electron to the adsorbed monomer.

Initiation by Strong Anions

Nucleophilic initiators include covalent or ionic metal amides, alkoxides, hydroxides, cyanides, phosphines, amines and organometallic compounds (alkyllithium compounds and Grignard reagents). The initiation process involves the addition of a neutral (B:) or negative (B:-) nucleophile to the monomer. The most commercially useful of these initiators has been the alkyllithium initiators. They are primarily used for the polymerization of styrenes and dienes.

Monomers activated by strong electronegative groups may be initiated even by weak anionic or neutral nucleophiles (i.e. amines, phosphines). Most prominent example is the curing of cyanoacrylate, which constitutes the basis for superglue. Here, only traces of basic impurities are sufficient to induce an anionic addition polymerization or zwitterionic addition polymerization, respectively.

Propagation

Propagation in anionic addition polymerization results in the complete consumption of monomer. This stage is often fast, even at low temperatures.

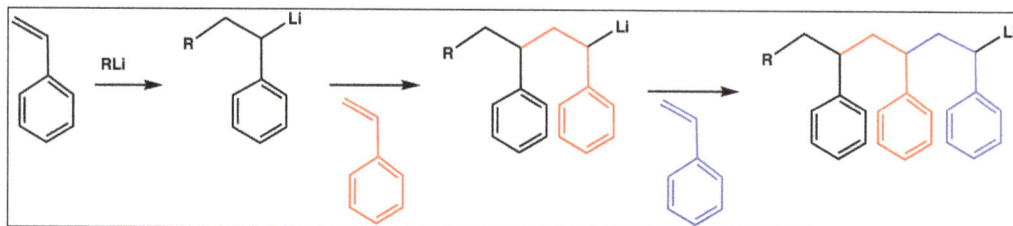

Organolithium-initiated polymerization of styrene.

Living Anionic Polymerization

Living anionic polymerization is a living polymerization technique involving an anionic propagating species.

Living anionic polymerization was demonstrated by Szwarc and co workers in 1956. Their initial work was based on the polymerization of styrene and dienes. One of the remarkable features of living anionic polymerization is that the mechanism involves no formal termination step. In the absence of impurities, the carbanion would still be active and capable of adding another monomer. The chains will remain active indefinitely unless there is inadvertent or deliberate termination or chain transfer. This gave rise to two important consequences:

- The number average molecular weight, M_n, of the polymer resulting from such a system could be calculated by the amount of consumed monomer and the initiator used for the polymerization, as the degree of polymerization would be the ratio of the moles of the monomer consumed to the moles of the initiator added.

  $M_n = M_o \dfrac{[M]_o}{[I]}$, where M_o = formula weight of the repeating unit, $[M]_o$ = initial concentration of the monomer, and $[I]$ = concentration of the initiator.

- All the chains are initiated at roughly the same time. The final result is that the polymer synthesis can be done in a much more controlled manner in terms of the molecular weight and molecular weight distribution (Poisson distribution).

The following experimental criteria have been proposed as a tool for identifying a system as living polymerization system.

- Polymerization until the monomer is completely consumed and until further monomer is added.

- Constant number of active centers or propagating species.

- Poisson distribution of molecular weight.

- Chain end functionalization can be carried out quantitatively.

However, in practice, even in the absence of terminating agents, the concentration of the living anions will reduce with time due to a decay mechanism termed as spontaneous termination.

Consequences of Living Polymerization

Block Copolymers

Synthesis of block copolymers is one of the most important applications of living polymerization as it offers the best control over structure. The nucleophilicity of the resulting carbanion will govern the order of monomer addition, as the monomer forming the less nucleophilic propagating species may inhibit the addition of the more nucleophilic monomer onto the chain. An extension of the above concept is the formation of tri-block copolymers where each step of such a sequence aims to prepare a block segment with predictable, known molecular weight and narrow molecular weight distribution without chain termination or transfer.

Sequential monomer addition is the dominant method, also this simple approach suffers some limitations. Moreover, this strategy, enables synthesis of linear block copolymer structures that are not accessible via sequential monomer addition. For common A-b-B structures, sequential block copolymerization gives access to well defined block copolymers only if the crossover reaction rate constant is significantly higher than the rate constant of the homopolymerization of the second monomer, i.e., $k_{AA} >> k_{BB}$.

End-group Functionalization/Termination

One of the remarkable features of living anionic polymerization is the absence of a formal termination step. In the absence of impurities, the carbanion would remains active, awaiting the addition of new monomer. Termination can occur through unintentional quenching by impurities, often present in trace amounts. Typical impurities include oxygen, carbon dioxide, or water. Termination intentionally allows the introduction of tailored end groups.

Living anionic polymerization allow the incorporation of functional end-groups, usually added to quench polymerization. End-groups that have been used in the functionalization of α-haloalkanes include hydroxide, $-NH_2$, -OH, -SH, -CHO, -COCH$_3$, -COOH, and epoxides.

An alternative approach for functionalizing end-groups is to begin polymerization with a functional anionic initiator. In this case, the functional groups are protected since the ends of the anionic polymer chain is a strong base. This method leads to polymers with controlled molecular weights and narrow molecular weight distributions.

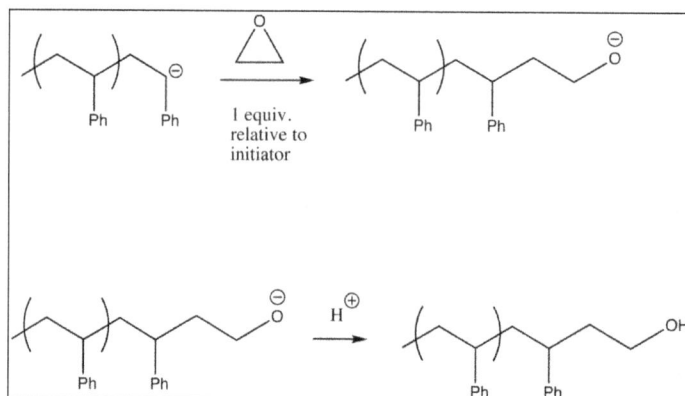

Addition of hydroxide group through an epoxide.

Ionic Polymerization

Ionic polymerization is a chain-growth polymerization in which active centers are ions or ion pairs. It can be considered as an alternative to radical polymerization, and may refer to anionic polymerization or cationic polymerization.

As with radical polymerization, reactions are initiated by a reactive compound. For cationic polymerization, titanium-, boron-, aluminum-, and tin-halide complexes with water, alcohols, or oxonium salts are useful as initiators, as well as strong acids and salts such as $KHSO_4$. Meanwhile, group 1 metals such as lithium, sodium, and potassium, and their organic compounds (e.g. sodium naphthalene) serve as effective anionic initiators. In both anionic and cationic polymerization, each charged chain end (negative and positive, respectively) is matched by a counterion of opposite charge that originates from the initiator. Because of the charge stability necessary in ionic polymerization, monomers which may be polymerized by this method are few compared to those available for free radical polymerization. Stable polymerizing cations are only possible using monomers with electron-releasing groups, and stable anions with monomers with electron-withdrawing groups as substituents.

While radical polymerization rate is governed nearly exclusively by monomer chemistry and radical stability, successful ionic polymerization is as strongly related to reaction conditions. Poor monomer purity quickly leads to early termination, and solvent polarity has a great effect on reaction rate. Loosely-coordinated and solvated ion pairs promote more reactive, fast-polymerizing chains, unencumbered by their counterions. Unfortunately, molecules that are polar enough to support these solvated ion pairs often interrupt the polymerization in other ways, such as by destroying propagating species or coordinating with initiator ions, and so they are seldom utilized. Typical solvents for ionic polymerization include non-polar molecules such as pentane, or moderately polar molecules such as chloroform.

Applications

Because of the polarity of the active group on each polymerizing radical, termination by chain combination is not seen in ionic polymerization. Furthermore, because charge propagation can only occur by covalent bond formation with the compatible monomer species, termination by chain transfer or disproportionation is impossible. This means that all polymerizing ions, unlike in radical polymerization, grow and maintain their chain lengths throughout the reaction duration (so-called "living" polymer chains), until termination by the addition of a terminating molecule such as water. This leads to virtually monodisperse polymer products, which have many applications in material analysis and product design. Furthermore, because the ions do not self-terminate, block copolymers may be formed by the addition of a new monomer species.

A few important uses of anionic polymerization include the following:

- Calibration standards for gel permeation chromatography.

- Microphase separating block copolymers.

- Thermoplastic elastomeric materials.

Ring-opening Metathesis Polymerization

Ring-opening metathesis polymerization (ROMP) uses metathesis catalysts to generate polymers from cyclic olefins. ROMP is most effective on strained cyclic olefins, because the relief of ring strain is a major driving force for the reaction – cyclooctene and norbornenes are excellent monomers for ROMP, but cyclohexene is very reluctant to form any significant amount of polymer. Norbornenes are favorite monomers for ROMP, as a wide range of monomer functionalities are easily available through Diels-Alder reactions.

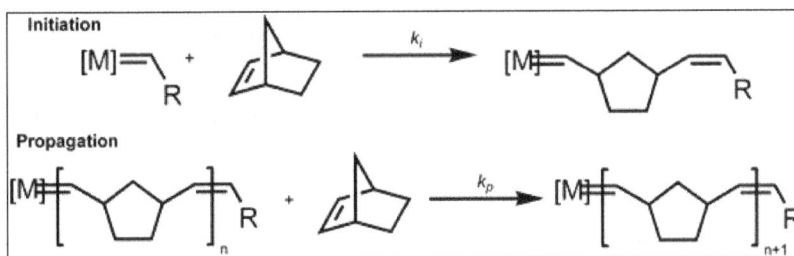

Careful balance of catalyst, monomer, and other factors can offer excellent control of the polymer structure. In terms of homogeneous catalysts, most tungsten and molybdenum catalysts (Schrock catalysts) have rapid initiation rates and can produce "living" polymerizations with excellent control of polydispersity and chain tacticity, but the low functional group tolerance limits the monomers available. Ruthenium metathesis catalysts (Grubbs catalysts) tend to have slower initiation rates, often leading to higher polydispersities, but their air stability and greater tolerance for functional groups makes them "user friendly" and enables use of a wide range of functional monomers and additives.

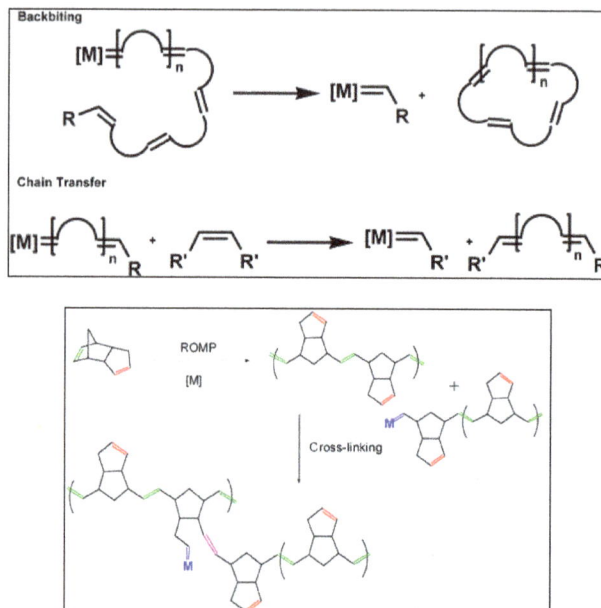

Secondary metathesis reactions (controlled by catalyst choice and reaction conditions) also affect the product distribution. Recoordination of an alkene on the growing polymer chain with the catalyst can lead to cyclic oligomers through a ring-closing metathesis reaction ("backbiting"). Chain transfer (cross metathesis) between a growing polymer unit and an adjacent polymer alkene also leads to broadened molecular weights. Chain transfer can also be used to improve processability of the resulting polymer – addition of an acyclic olefin (chain-transfer agent) can limit chain molecular weights and introduce terminal functional groups.

References

- Polymerization, science: britannica.com, Retrieved 28 July, 2019

- Tsutomu Yokozawa; Yoshihiro Ohta (2016). "Transformation of Step-Growth Polymerization into Living Chain-Growth Polymerization". Chemical Reviews. 116 (4): 1950–1968. doi:10.1021/acs.chemrev.5b00393. PMID 26555044

- Ring-opening-metathesis-polymerization: allthingsmetathesis.com, Retrieved 17 January, 2019

- Jenkins AD, Jones RG, Moad G (2009). "Terminology for reversible-deactivation radical polymerization previously called "controlled" radical or "living" radical polymerization (IUPAC Recommendations 2010)". Pure and Applied Chemistry. 82 (2): 483–491. doi:10.1351/PAC-REP-08-04-03. ISSN 1365-3075

- Fahlman, Bradley D. (2008). Materials Chemistry. Springer. ISBN 978-1-4020-6119-6

- R. Auras; L.-T. Lim; S. E. M. Selke; H. Tsuji (2010). Poly(lactic acid): Synthesis, Structures, Properties, Processing, and Applications. Wiley. ISBN 978-0-470-29366-9

5

Polymers Stereochemistry

> The subdiscipline of chemistry which deals with the study of the relative spatial arrangement of atoms which form the structure of molecules and their manipulation is known as stereochemistry. This chapter closely examines the key concepts of stereochemistry of polymers to provide an extensive understanding of the subject.

When a polymer has stereochemical isomerism within the chain, its properties often depend on the stereochemical structure.

Symmetrical monomers such as ethylene and tetrafluoroethylene can join together in only one way. Monosubstituted monomers, on the other hand, may join together in two organized ways, described in the following diagram, or in a third random manner. Most monomers of this kind, including propylene, vinyl chloride, styrene, acrylonitrile and acrylic esters, prefer to join in a head-to-tail fashion, with some randomness occurring from time to time.

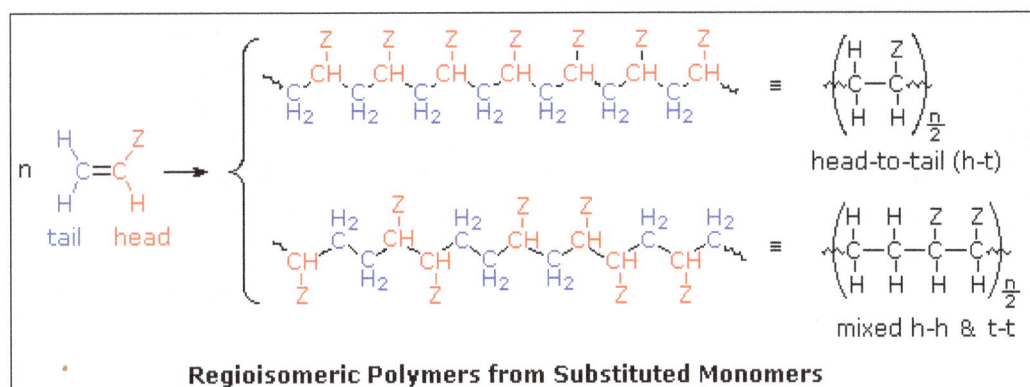

Regioisomeric Polymers from Substituted Monomers

If the polymer chain is drawn in a zig-zag fashion, as shown above, each of the substituent groups (Z) will necessarily be located above or below the plane defined by the carbon chain. Consequently we can identify three configurational isomers of such polymers. If all the substituents lie on one side of the chain the configuration is called isotactic. If the substituents alternate from one side to another in a regular manner the configuration is termed syndiotactic. Finally, a random arrangement of substituent groups is referred to as atactic. Examples of these configurations are shown here.

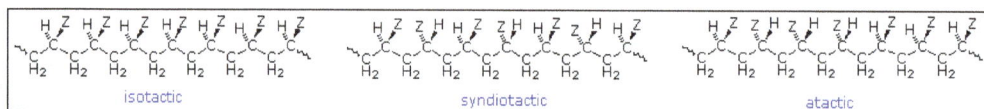

isotactic   syndiotactic   atactic

Many common and useful polymers, such as polystyrene, polyacrylonitrile and poly(vinyl chloride) are atactic as normally prepared. Customized catalysts that effect stereoregular polymerization of polypropylene and some other monomers have been developed, and the improved properties associated with the increased crystallinity of these products has made this an important field of investigation. The following values of Tg have been reported.

Polymer	T_g atactic	T_g isotactic	T_g syndiotactic
PP	−20 °C	0 °C	−8 °C
PMMA	100 °C	130 °C	120 °C

The properties of a given polymer will vary considerably with its tacticity. Thus, atactic polypropylene is useless as a solid construction material, and is employed mainly as a component of adhesives or as a soft matrix for composite materials. In contrast, isotactic polypropylene is a high-melting solid (ca. 170 °C) which can be molded or machined into structural components.

Stereoisomerism of Polymers

Stereoisomerism is a final type of isomeric variation occurs as a result of the three-dimensional structure of some polymers. It is possible because a four-valent atom like carbon can exist in two different forms when the subsidiary groups or atoms attached to the carbon are all different. The carbon atom is then known as an asymmetric carbon atom. A very simple example of the phenomenon is the structure of a small molecule, lactic acid. As figure shows, it can exist in two forms which are mirror images of one another. One of the two possible compounds, laevo- (standing for left-handed), or l-lactic acid occurs in muscle after vigorous, anaerobic exercise and causes muscle cramp. It is a good example of stereoisomerism in a small molecule, a feature it also shares with a large number of biological molecules, as well as some polymers.

The carbon atom in a vinyl polymer to which is attached the pendant side group (i.e. every alternate carbon atom in the main chain) is another example of an asymmetric carbon atom. It gives rise to tacticity. When the zig-zag chain is written with the plane of the chain in the plane of the paper, there are three ways in which the position of the pendant side group can exist. The methyl groups in polypropylene, for example, can occur all on one side (isotactic), on alternate sides (syndiotactic) or placed at random (atactic). These possibilities are shown in figure (c). The properties of each type of polymer are quite different to one another, primarily because isotactic and syndiotactic PP have ordered chains and so can crystallise, but atactic chains are quite irregular and cannot crystallise. Isotactic PP is the common form of the commercial material, although atactic PP is used as a binder for paper for example. Syndiotactic PP has recently become available commercially being made using a new family of catalysts, known as metallocenes.

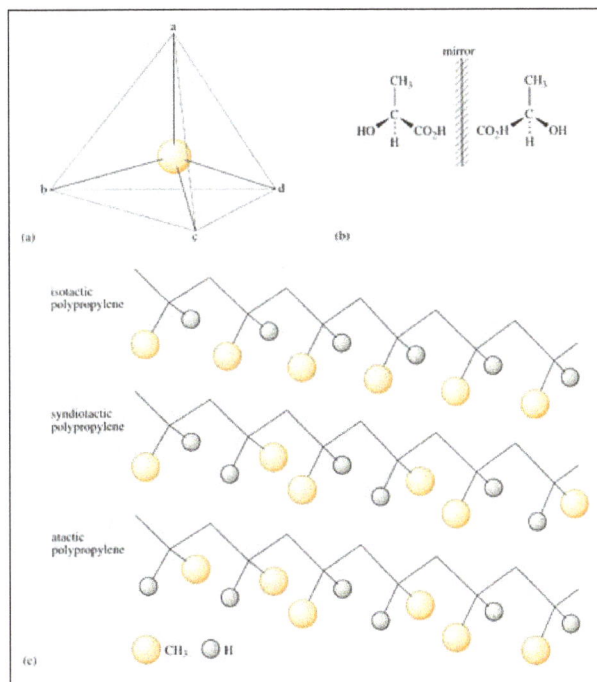

(a) Tetrahedral configuration of single carbon atom; (b) left and right-handed forms of lactic acid; (c) tacticity in polypropylene.

Tacticity

Tacticity is the relative stereochemistry of adjacent chiral centers centers within a macromolecule. The practical significance of tacticity rests on the effects on the physical properties of the polymer. The regularity of the macromolecular structure influences the degree to which it has rigid, crystalline long range order or flexible, amorphous long range disorder. Precise knowledge of tacticity of a polymer also helps understanding at what temperature a polymer melts, how soluble it is in a solvent and its mechanical properties.

A tactic macromolecule in the IUPAC definition is a macromolecule in which essentially all the configurational (repeating) units are identical. Tacticity is particularly significant in vinyl polymers of the type $-H_2C-CH(R)-$ where each repeating unit with a substituent R on one side of the polymer backbone is followed by the next repeating unit with the substituent on the same side as the previous one, the other side as the previous one or positioned randomly with respect to the previous one. In a hydrocarbon macromolecule with all carbon atoms making up the backbone in a tetrahedral molecular geometry, the zigzag backbone is in the paper plane with the substituents either sticking out of the paper or retreating into the paper. This projection is called the Natta projection after Giulio Natta. Monotactic macromolecules have one stereoisomeric atom per repeat unit, ditactic to n-tactic macromolecules have more than one stereoisomeric atom per unit.

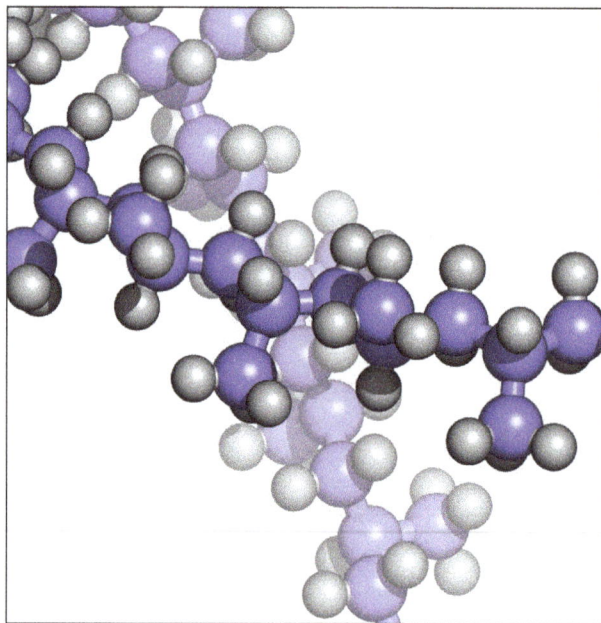

A ball-and-stick model of syndiotactic polypropylene.

Describing Tacticity

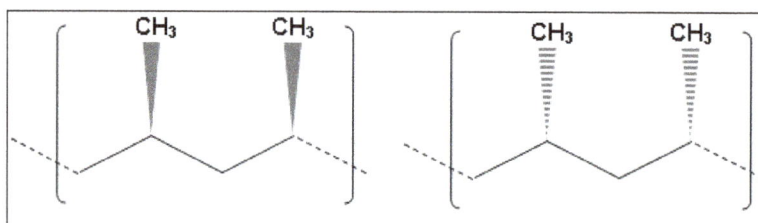

An example of *meso* diads in a polypropylene molecule.

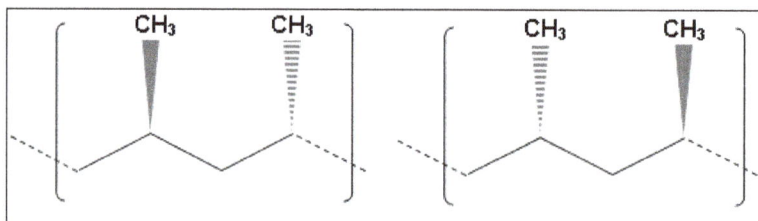

An example of *racemo* diads in a polypropylene molecule.

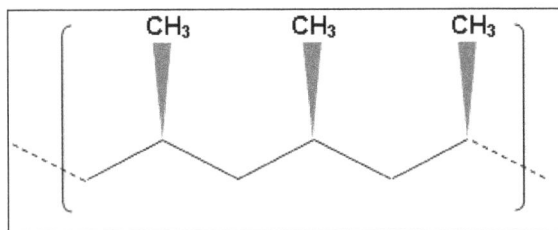

An isotactic (*mm*) triad in a polypropylene molecule.

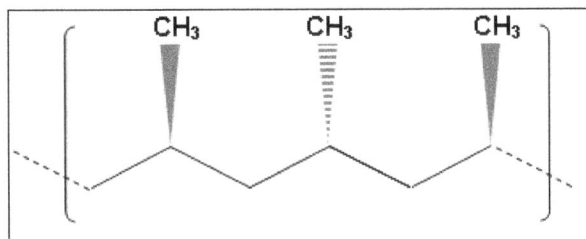

A syndiotactic (*rr*) triad in a polypropylene molecule.

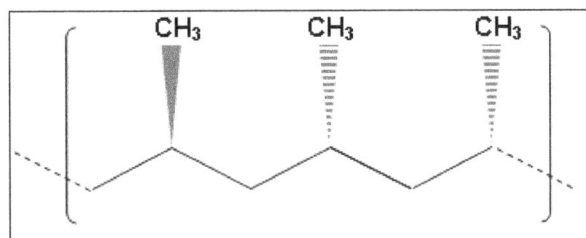

A heterotactic (*rm*) triad in a polypropylene molecule.

Diads

Two adjacent structural units in a polymer molecule constitute a diad. If the diad consists of two identically oriented units, the diad is called a meso diad reflecting similar features as a meso compound. If the diad consists of units oriented in opposition, the diad is called a racemo diad as in a racemic compound. In the case of vinyl polymer molecules, a meso diad is one in which the book carbon chains are oriented on the same side of the polymer backbone.

Triads

The stereochemistry of macromolecules can be defined even more precisely with the introduction of triads. An isotactic triad (mm) is made up of two adjacent meso diads, a syndiotactic triad (also spelled syndyotactic) {rr} consists of two adjacent racemo diads and a heterotactic triad (rm) is composed of a meso diad adjacent to a racemo diad. The mass fraction of isotactic (mm) triads is a common quantitative measure of tacticity.

When the stereochemistry of a macromolecule is considered to be a Bernoulli process, triad composition can be calculated from the probability of finding meso diads (P_m). When this probability is 0.25 then the probability of finding:

- An isotactic triad is P_m^2 or 0.0625.

- An heterotactic triad is $2P_m(1-P_m)$ or 0.375.

- A syndiotactic triad is $(1-P_m)^2$ or 0.5625.

with a total probability of 1. Similar relationships with diads exist for tetrads.

Tetrads and Pentads

The definition of tetrads and pentads introduce further sophistication and precision to defining tacticity, especially when information on long-range ordering is desirable. Tacticity measurements obtained by Carbon-13 NMR are typically expressed in terms of the relative abundance of various pentads within the polymer molecule, e.g. *mmmm*, *mrrm*.

Other Conventions for Quantifying Tacticity

The primary convention for expressing tacticity is in terms of the relative weight fraction of triad or higher-order components, as described above. An alternative expression for tacticity is the average length of *meso* and *racemo* sequences within the polymer molecule. The average meso sequence length may be approximated from the relative abundance of pentads as follows:

$$MSL = \frac{mmmm + \frac{3}{2}mrrr + 2rmmr + \frac{1}{2}rmrm + \frac{1}{2}rmrr}{\frac{1}{2}mmmr + rmmr + \frac{1}{2}rmrm + \frac{1}{2}rmrr}$$

Isotactic Polymers

Isotactic polymers are composed of isotactic macromolecules (IUPAC definition). In isotactic macromolecules all the substituents are located on the same side of the macromolecular backbone. An isotactic macromolecule consists of 100% meso diads. Polypropylene formed by Ziegler–Natta catalysis is an isotactic polymer. Isotactic polymers are usually semicrystalline and often form a helix configuration.

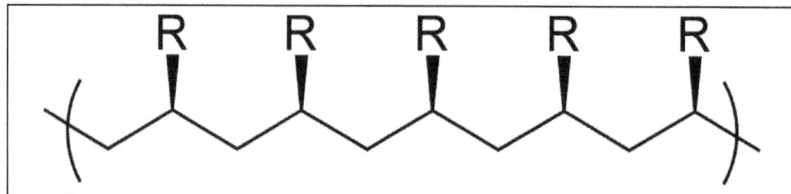

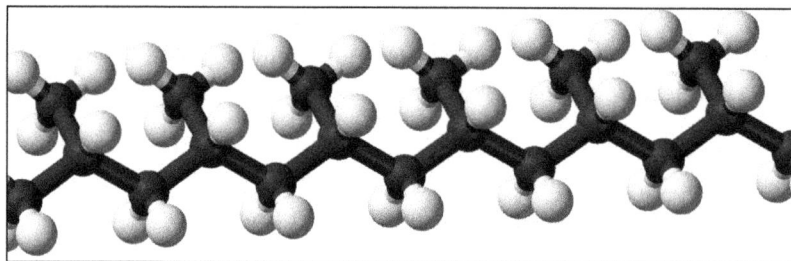

Syndiotactic Polymers

In syndiotactic or syntactic macromolecules the substituents have alternate positions along the chain. The macromolecule consists 100% of racemo diads. Syndiotactic polystyrene, made by metallocene catalysis polymerization, is crystalline with a melting point of 161 °C. Gutta percha is also an example for Syndiotactic polymer.

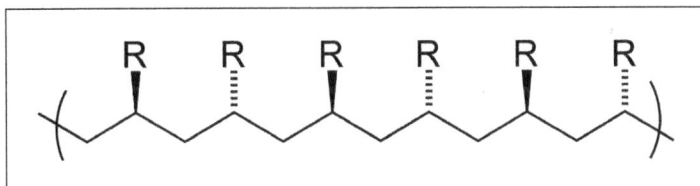

Atactic Polymers

In atactic macromolecules the substituents are placed randomly along the chain. The percentage of meso diads is between 1 and 99%. With the aid of spectroscopic techniques such as NMR it is possible to pinpoint the composition of a polymer in terms of the percentages for each triad.

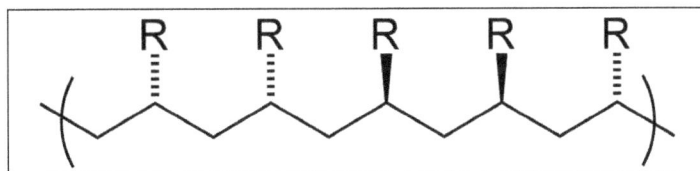

Polymers that are formed by free-radical mechanisms such as polyvinyl chloride are usually atactic. Due to their random nature atactic polymers are usually amorphous. In hemi isotactic macromolecules every other repeat unit has a random substituent.

Atactic polymers are technologically very important. A good example is polystyrene (PS). If a special catalyst is used in its synthesis it is possible to obtain the syndiotactic version of this polymer, but most industrial polystyrene produced is atactic. The two materials have very different properties because the irregular structure of the atactic version makes it impossible for the polymer chains to stack in a regular fashion. The result is that, whereas syndiotactic PS is a semicrystalline material, the more common atactic version cannot crystallize and forms a *glass* instead. This example is quite general in that many polymers of economic importance are atactic glass formers.

Eutactic Polymers

In eutactic macromolecules, substituents may occupy any specific (but potentially complex) sequence of positions along the chain. Isotactic and syndiotactic polymers are instances of the more general class of eutactic polymers, which also includes heterogeneous macromolecules in which the sequence consists of substituents of different kinds (for example, the side-chains in proteins and the bases in nucleic acids).

Head/Tail Configuration

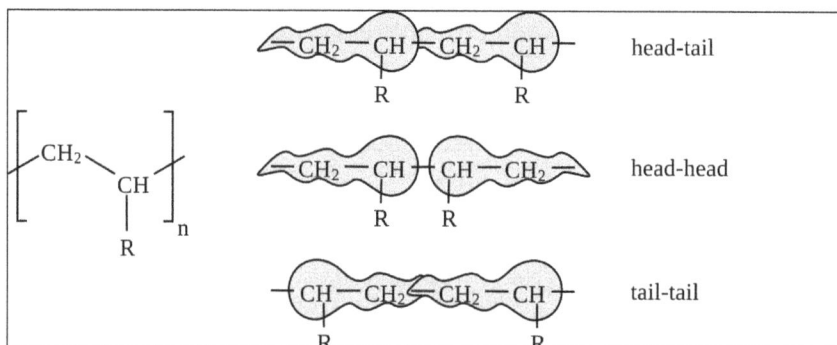

In vinyl polymers the complete configuration can be further described by defining polymer head/tail configuration. In a regular macromolecule all monomer units are normally linked in a head to tail configuration so that all β-substituents are separated by three carbon atoms. In head to head configuration this separation is only by 2 carbon atoms and the separation with tail to tail configuration is by 4 atoms. Head/tail configurations are not part of polymer tacticity but should be taken into account when considering polymer defects.

Techniques for Measuring Tacticity

Tacticity may be measured directly using proton or carbon-13 NMR. This technique enables quantification of the tacticity distribution by comparison of peak areas or integral ranges corresponding to known diads (r, m), triads (mm, rm+mr, rr) and/or higher order n-ads depending on spectral resolution. In cases of limited resolution stochastic methods such as Bernoullian or Markovian analysis may also be used to fit the distribution and back then forward predict higher n-ads and calculate the isotacticity of the polymer to the desired level.

Other techniques sensitive to tacticity include x-ray powder diffraction, secondary ion mass spectrometry (SIMS), vibrational spectroscopy (FTIR) and especially two-dimensional techniques. Tacticity may also be inferred by measuring another physical property, such as melting temperature, when the relationship between tacticity and that property is well-established.

Steric Arrangement in Cis and Trans Configuration

Polymers with double bonds in the repeat unit give rise to different geometric isomers. For example, 1,3-butadiene can be polymerized to give poly(1,2-butadiene) or either of two geometric isomers of poly(1,4-butadiene). These two isomers are called *cis* and *trans* poly(1,4-butadiene). In the case of cis-polybutadiene, the first and the forth carbon lie on the same side of the central double bond, and in the case of trans, on opposite sides of the central double bond.

cis-1,4-polyisoprene   trans-1,4-polyisoprene

Double bonds in the cis form reduce the energy barrier for the rotation of adjacent C-C bonds compared to those in trans from, which, in turn, reduces the glass transition temperature (T_g). The affect of cis-trans isomerie on the T_g is usually not very strong, that is, the T_g shift is only about 20 K for polybutadiene. If other substituents (-Cl, -CH_3 etc.) are present, the energy barrier for the rotation of adjacent bonds will increase, and hence the T_g will increase as well. Bulky substituents will also reduce the energy barrier for rotation of the trans form relative to the cis form. For example, the observed T_gs of cis and trans polyisoprene are nearly the same, whereas the T_g of the cis isomer of poly(chloro-butadiene) is about 20 K larger than the T_g of the trans isomer.

Glass Transition Temperature of Polybutadienes		
Polymer	$T_{g,cis}$ (K)	$T_{g,trans}$ (K)
Polybutadiene	170 (165)	187 (190)
Polyisoprene	204	207
Poly(chlorobutadiene)	252	234

The double bonds in the polymer backbone prevent rotation that occurs in carbon-carbon single bonds. It is believed that the inability to rotate freely hinders the allignment of the polymer chains. Thus, unsaturation in the polymer chain reduces the crystallinity and increases the elasticity and flexibility of polymers.

Permissions

All chapters in this book are published with permission under the Creative Commons Attribution Share Alike License or equivalent. Every chapter published in this book has been scrutinized by our experts. Their significance has been extensively debated. The topics covered herein carry significant information for a comprehensive understanding. They may even be implemented as practical applications or may be referred to as a beginning point for further studies.

We would like to thank the editorial team for lending their expertise to make the book truly unique. They have played a crucial role in the development of this book. Without their invaluable contributions this book wouldn't have been possible. They have made vital efforts to compile up to date information on the varied aspects of this subject to make this book a valuable addition to the collection of many professionals and students.

This book was conceptualized with the vision of imparting up-to-date and integrated information in this field. To ensure the same, a matchless editorial board was set up. Every individual on the board went through rigorous rounds of assessment to prove their worth. After which they invested a large part of their time researching and compiling the most relevant data for our readers.

The editorial board has been involved in producing this book since its inception. They have spent rigorous hours researching and exploring the diverse topics which have resulted in the successful publishing of this book. They have passed on their knowledge of decades through this book. To expedite this challenging task, the publisher supported the team at every step. A small team of assistant editors was also appointed to further simplify the editing procedure and attain best results for the readers.

Apart from the editorial board, the designing team has also invested a significant amount of their time in understanding the subject and creating the most relevant covers. They scrutinized every image to scout for the most suitable representation of the subject and create an appropriate cover for the book.

The publishing team has been an ardent support to the editorial, designing and production team. Their endless efforts to recruit the best for this project, has resulted in the accomplishment of this book. They are a veteran in the field of academics and their pool of knowledge is as vast as their experience in printing. Their expertise and guidance has proved useful at every step. Their uncompromising quality standards have made this book an exceptional effort. Their encouragement from time to time has been an inspiration for everyone.

The publisher and the editorial board hope that this book will prove to be a valuable piece of knowledge for students, practitioners and scholars across the globe.

Index

www.ingramcontent.com/pod-product-compliance
Lightning Source LLC
Chambersburg PA
CBHW062002190326
41458CB00009B/2939

* 9 7 8 1 6 4 1 7 2 3 7 3 2 *